CHEMISTRY OF THE
SOLAR SYSTEM

CHEMISTRY OF THE SOLAR SYSTEM

An Elementary Introduction To Cosmochemistry

HANS E. SUESS

Chemistry Department
University of California, San Diego
La Jolla, California

A Wiley-Interscience Publication

JOHN WILEY & SONS

New York   Chichester   Brisbane   Toronto   Singapore

Copyright © 1987 by John Wiley & Sons, Inc.

All rights reserved. Published simultaneously in Canada.

Reproduction or translation of any part of this work
beyond that permitted by Section 107 or 108 of the
1976 United States Copyright Act without the permission
of the copyright owner is unlawful. Requests for
permission or further information should be addressed to
the Permissions Department, John Wiley & Sons, Inc.

Library of Congress Cataloging in Publication Data:

Suess, Hans Eduard, 1909–
   Chemistry of the solar system.
   Bibliography: p.
   Includes index.
   1. Cosmochemistry. I. Title.
QB450.S48 1987   523.02   86-24644
ISBN 0-471-83107-7

Printed in the United States of America

10 9 8 7 6 5 4 3 2 1

Preface

During the years 1968 to 1985, I taught a course entitled "Introduction to Cosmochemistry" at the University of California, San Diego. The course was intended primarily for seniors graduating in chemistry and earth sciences. Occasionally biologists, as well as students majoring in English literature and history, attended. Some of them seemed to wish to acquire some background for science fiction writing. The course was not taught for astronomy and physics majors, as students in these fields look at this subject from a different perspective. I was trying to convince the students that we do have some knowledge of the events and processes that occurred before our planet Earth existed. In some areas, this knowledge is firmly established and reliable; in other areas, we must be satisfied with hypothetical conjectures. Most students do not realize that it is necessary to distinguish between these two basically different situations.

One reason for writing this book is that many students asked me where they could read more about what I was teaching. The material that I could recommend, however, was scattered throughout the literature and contained either too much or too little information. Another reason for this book is that my own contributions, many of them accomplished in cooperation with Hans Jensen, have been published over the years in the form of brief notes in various journals.

I have tried to outline some elementary background, that may be well-known to most students of chemistry, although not in regard to its relationship to Earth and planetary sciences. In many cases, I have tried to define interesting questions in the field without discussing hypothetical answers. The starting point of the approach to cosmochemical problems that I outline is the chemical composition of the matter from which our solar system formed. The main point of the book is to convince the student that we know this composition accurately and reliably, indeed, more accurately and reliably than many other facts and data on which our knowledge of the origin and development of the solar system and its members is based. To show this, I have given an extended review of elementary nuclear physics, specifically covering the topics needed to understand the arguments demonstrating the validity of elemental and nuclear abundance values that must be used to derive further conclusions. Much of what is said here can be found in many textbooks, but it is included for the benefit of those who have little background in these areas. Students of cosmochemistry should also be familiar with topics that so far are discussed only in original publications. These topics are discussed here in detail. Although the text is intended to be rather elementary, my colleagues who read this will, here and there, find statements that are contrary to current views and represent new ideas. I could not help but include these, as I feel strongly that mainstream opinion is sometimes on the wrong track.

HANS E. SUESS

La Jolla, California
February 1987

Acknowledgments

I thank my friends and colleagues who have read this text and have made valuable comments, among them Professors Ed Goldberg, James Arnold, Günter Lugmeyer, Klaus Keil, Kurt Marti, Mark Thiemen, Dieter Zeh, and also my former students Drs. Candace Kohl, Gerlind Dreibus, and Rick Fox. Typing of the manuscript was done by Teresa Jackson, as well as by Dharm Darshan, who also did the first careful editing. The last section on planets was in part written in collaboration with Professor Heinrich Wänke of the Max-Planck-Institut in Mainz, West Germany. Financial support for this collaboration came in part from the Alexander von Humboldt Foundation. Also, I wish to express my appreciation to NASA for a grant that has supported more than a decade of my work on the abundances of the elements.

Finally, above all I am greatly indebted to Professor Harrison Brown, not only for his perseverance in overcoming time-consuming formalities that allowed me to accept an invitation from the University of Chicago in 1950, but also for many illuminating and encouraging discussions on our joint fields of science.

H.E.S.

Contents

Introduction

The field of science dealing with the behavior of the chemical elements on the surface in the interior of the Earth is commonly defined as geochemistry. Primarily, it deals with the question of how the chemical elements behave during geological processes. One may also ask how the chemical elements behave in our solar system as a whole. This question belongs to the more general field now called "cosmochemistry." Hence, geochemistry should be considered a subfield of cosmochemistry. Cosmochemistry is related to astrophysics, which is concerned with the reactions in the interior of stars. Because of the high temperatures prevailing there, these are nuclear reactions.

The field of cosmochemistry can be limited to the question of how the chemical elements, and in particular their isotopes, behaved during and after the formation of the solar system. This behavior determined the chemical composition of the members of our solar system—the planets, their satellites, and the smaller objects, the asteroids, comets, meteorites, and so on.

Why do we want to know this? Certainly we want to know how the chemical composition of the surface of the Earth evolved and how it changed during geologic time. In this connection it would surely be very useful to know how conditions on the Earth's surface compare with those on other planets. The goal of the NASA planetary program is to learn about this, partly to understand what happened on our own planet during geologic time. The most exciting question concerns the evolution of life. The Earth is a peculiar planet, with a unique history. Conditions on our planet's surface are different from those on other planets. Why is this so? What conditions are necessary for the evolution of life? A fundamental scientific question is: How probable is it that life develops under the conditions that have prevailed on Earth? It may be that under certain chemical and physical conditions life always evolves. It may also be that the evolution of life as it exists on Earth is exceedingly improbable. It may have occurred in the whole universe only once, here on our Earth. The answer to this question will affect our thinking in many ways. In order to learn about the evolution of life on Earth, we need to know about conditions on the surface of the early, primitive Earth. The field of cosmochemistry addresses this question.

It is important to remember that our understanding of the solar system and the stars was not acquired out of intellectual curiosity. This knowledge came about first with the advent of food production nearly 10,000 years ago, when it became necessary to predict seasonal changes to time planting and harvesting and the care of domesticated animals. The neolithic site of Stonehenge, which dates from the second millenium B.C., is thought to represent a prehistoric observatory that was used to determine the number of days it takes the Sun to return to the same place when it sets on the horizon and the number of days that elapse from one full Moon to the next. Long before historic records were available, our ancestors must have known not only about the Sun and Moon, but also about the stars, which appeared permanently fixed on a sphere revolving around the Earth, except for the five luminous bodies that moved about the others in an apparently irregular manner.

We humans, and probably higher animals, distinguish between unanimated and animated objects by assuming that those objects that behave in a strictly predictable manner are unanimated and those that behave in an unpredictable way are animated. Hence from very early days, probably simultaneously with the development of speech and language, people attributed the irregular motions of the planets to the whims of supernatural beings, gods, and spirits, who affected events and lives on Earth. This belief led to the development of astrology, the study of motions of the planets in the sky, in order to predict the future. Astrology is more than 5000 years old. It developed in the Near East during the Sumerian and Babylonian empires. Early names of the five visible planets, together with the Sun and Moon, are still with us in our everyday lives. Apparently the Babylonians were the first to name the seven days of the week after the Sun, Moon and five planets, and the Romans later adapted these names. Somewhat modified, they survived the transition into Christianity. Table I shows that they can still be recognized in Latin languages. In the Germanic languages, these names were replaced by those of the corresponding Teutonic gods.

During the preceding 3000–4000 years, accurate observations of the movements of the planets in the sky led to acceptance of a consistent scheme that depicted the Sun, Moon, and planets as revolving around the Earth; this view was accepted until the late Middle Ages. In the fourth century B.C., the Greek philosopher Aristotle postulated a universe with the Earth as a flat disk surrounded by the heavenly bodies. This cosmology was accepted by the Holy Roman Church. In the 15th century A.D., people contemplated another old and forgotten, though more logical, model: the Earth as a sphere revolving about the Sun in the center. Nicolaus Copernicus (1473–1543), then Galileo Galilei (1564–1642), and, in a more quantitative way, Johannes Kepler (1571–1630) held that the planets revolved in nearly circular, elliptical orbits at regular distances from the Sun with mathematically defined periods.

One would think this would not be of any great importance to the general public. Why should people get excited about such questions? Why should mighty rulers of countries and of the Christian Church be enraged by these ideas? I am sure that even today the majority of people do not much care about whether the Earth goes around the Sun or the Sun around the Earth. Why should money be spent on such a question that has nothing to do with our standard of living, defense, or lifestyle? Yet it might be said that this question of whether the Sun revolves around the Earth or the Earth around the Sun, asked some 300 years ago, contributed decisively to changes in the whole political structure of the Western world. We know that when people were starting to argue about this question, the political forces and religious powers of Western society were strongly opposed to such ideas. Those who claimed that the Earth revolved around the Sun were imprisoned or burned at the stake. Scientists who claimed that the Earth revolved around the sun did not themselves understand why this belief was considered religious and political heresy. Yet the implications were profound, as the whole social structure at that time was based on the idea that the Earth was the center of the universe.

The question of whether the evolution of life is a probable or an

TABLE I
Names of Weekdays, Gods, and Planets

	Weekdays				Gods		Celestial Bodies
English	German	French	Latin	Germanic	Roman		
Sunday	Sonntag	Dimanche	Dies Solis				Sun
Monday	Montag	Lundi	Lunae Dies				Moon
Tuesday	Dienstag	Mardi	Martis Dies	Zeus, Ziu	Mars		Mars
Wednesday	Mittwoch	Mercredi	Dies Mercurii	Wodan, Odin	Mercur		Mercury
Thursday	Donnerstag	Jeudi	Jovis Dies	Donar, Thor	Jupiter		Jupiter
Friday	Freitag	Vendredi	Veneris Dies	Fria	Venus		Venus
Saturday	Samstag	Samedi	Saturni Dies		Saturn		Saturn

improbable event in the universe may be of similar importance today. We do not have the answer to this question yet. Future research will certainly tell us. We may find that life always evolves on a planet with a history similar to that of our Earth, or we may find that the evolution of life as it exists on our planet is a rare, unique event. Knowing this would certainly influence our thinking in many ways, and its impact may be as great as that of the ideas of Copernicus and Kepler. The Earth may well be the only planet inhabited by thinking, observing organisms. Planet Earth is certainly unique, and not only in our solar system. We must expect that its chemical and physical conditions are relatively rare in our galaxy and in the universe. However, even though an Earth-like planet may be present only once in millions of planetary systems, an enormous number of Earth-like planets probably exist in the universe as a whole.

The later part of the 19th century marked the beginning of serious attempts to address scientific questions of a chemical nature. On October 26, 1889, Frank Wigglesworth Clarke read a paper before the Philosophical Society of Washington entitled "The Relative Abundance of the Chemical Elements." It was the first of numerous articles by many different authors on this topic (Clarke, 1889). It contained the following statement: "An attempt was made in the course of this investigation to represent the relative abundances of the elements by a curve, taking their atomic weight for one set of ordinates. It was hoped that some sort of periodicity might be evident, but no such regularity appeared."

During the subsequent 50 years, Clarke and his coworker, H. S. Washington, continued this line of research at the U. S. Geological Survey. Their work is still considered to be one of the most valuable sources of geochemical facts. One observation based on the work of Clarke and Washington, now known as Harkins's rule, proved to be of fundamental importance. It states that elements with even atomic numbers are more abundant in nature than those with odd numbers. Clarke and Washington had concentrated their work on the composition of the Earth's crust. In time, however, it became increasingly evident that meteorites were better objects for the study of the average abundance of the chemical elements in nature than were terrestrial rocks. These studies culminated in Goldschmidt's classic paper of 1937, which has served as the basis of practically all subsequent work in this field.

When, in 1889, Clarke was attempting to uncover periodicities in the relative abundance of the elements, he expected to find some connection with the Periodic Table. Increased knowledge of the abundances of the elements, the discovery of isotopes, and the determinations of the isotopic composition of the elements led to substantial progress. It was possible more than 40 years later to detect certain types of periodicities that, however, followed laws different from those of atomic structure and were unrelated to the Periodic Table. An entirely new aspect began to reveal itself, promising to lead far deeper into fundamental questions of nature than Clarke had expected.

About 60 years after Clarke had presented his classical paper, Suess (1947) showed that there was a correlation between the relative abundances of the elements and their isotopic composition. This correlation conclusively demonstrated that the observed values for the abundances were largely mediated by their nuclear properties. The relative abundances of the individual nuclear species were connected in an irrefutable, though then inexplicable, way with nuclear properties. This meant that the matter surrounding us bore signs of representing the ashes of the nuclear fires that had led to its creation. Since then, much progress has been made toward understanding these reactions. In particular, it is now clear that this primeval matter consisted of several components that had undergone different nuclear histories. The nature of at least one of these components is now, in principle, well understood, but much concerning the origin of the chemical elements in our solar system is still a matter of surmise.

In any case, enough is known to allow us to postulate values for the average chemical composition of our solar system and to give values for the relative abundances of all the naturally occurring elements of the Periodic Table. The study of the nuclear and chemical processes involved in the formation of the members of the solar system is the essential subject matter of the field of cosmochemistry. Students of this field want to know how sure we can be that the members of our solar system evolved from a well-defined and almost entirely chemically homogeneous mass. And they want to know about the chemical reactions that had led to their formation. The discussion that follows attempts to provide the necessary background for an understanding of the answers to such questions.

We proceed here in a way typical for a chemist who wants to investigate the properties of an object. A chemist first carries out a qualitative analysis to determine what chemical elements are present, and thereafter a quantitative analysis to see how much of each chemical element is present. In our case, the object is our solar system. We then discuss the chemical state of the elements and how they might have been separated from each other to form the small and large members of the solar system. From this perspective, some typical cosmochemical problems are discussed here in detail.

PART ONE

NUCLEAR IMPLICATIONS

ONE

Elemental Analysis of the Solar System

1.1 QUALITATIVE ANALYSIS

1.1.1 The Elements and Their Isotopes

What is the result of a qualitative analysis of the solar system as a whole? The answer to this unusual question is very simple: Our sample contains the stable elements on the Periodic Table, that is, all the elements from hydrogen to uranium. The amounts in which they are present, however, vary enormously. For example, although the elements between bismuth and uranium are present, they are radioactive and exist in only extremely small amounts. Elements 43 and 61 are missing in terrestrial material, because all their isotopes are radioactive and have half-lives that are short compared to the age of the Earth. They may be present, however, as intermediate products of nuclear transformations that appear to occur on the surface of certain types of stars. The reasons why all the isotopes of these two elements are radioactive are usually not discussed in textbooks, but they will be considered here.

The Periodic Table of the elements was established simultaneously and independently in about 1870 by a Russian, Dimitri Ivan Mendeleev, and a German, Lothar Mayer. These investigators noticed that when the known elements were listed in sequence of their atomic weights, their chemical properties, and many of their physical properties, recurred in a periodic way. This is now fully understood: When the elements are numbered in sequence according to their atomic weight, the numbers simply denote the number of electrons associated with the respective atomic nucleus. The number of electrons determines the physical and chemical properties of an element.

An interesting example of how scientific progress occurs is the hypothesis of Prout. As early as 1815, Prout recognized that many elements have atomic weights remarkably close to integer multiples of the atomic weight of hydrogen. He discussed the possibility that the heavy elements consist of some kind of a polymer of hydrogen. Although Prout himself always emphasized that this relationship in atomic weights might be accidental, many researchers took it seriously, and great efforts were made to measure atomic weights as accurately as possible by determining the weight ratios in chemical compounds. Indeed, many elements such as carbon, nitrogen, and oxygen were found to have atomic weights very close to exact integer multiples of hydrogen. However, chlorine was clearly shown to deviate, as its atomic weight was experimentally observed to be between 35 and 36 times that of hydrogen.

Strangely, the more measurements were made, the more closely the atomic weights of several elements moved toward values of whole multiples of hydrogen, whereas those of others continued to deviate from such values. The suspicion therefore was justified, even before any experimental proof existed, that some of the elements were mixtures of atomic species that were chemically identical but had different atomic weights that were whole multiples of hydrogen. The final proof came in 1906, when J. J. Thomson, in England, constructed a primitive mass spectrograph, a device with which he could weigh individual atoms by observing the deflection of ion beams in a magnetic field. A moving charge is deflected when it passes through such a field, and the degree of deflection depends on the speed and mass of the particle. A photographic plate was used as a detector for the ions. In this way, different lines were obtained for the different masses of the individual atomic species. The mass differences corresponded closely to the mass of one or more hydrogen atoms. Thus, atoms of one and the same element were found to have different masses. Such atoms were called isotopes of the respective element.

In the years during and after World War I, Aston (1933), also in England, systematically investigated the isotopic composition of all the elements of the Periodic Table with greatly improved precision and accuracy. This led to another important discovery: Although individual isotopes of all the elements had masses close to an integer multiple of hydrogen, their values were slightly smaller. This was then correctly attributed to the so-called packing fraction, or mass defect, equal to the mass equivalent of the energy $E = mc^2$ released when individual hydrogen atoms (protons plus electrons) were combined to form a heavy atom. The values of these mass defects were then accurately determined by Dempster and by Bainbridge in the United States in 1931 and by Mattauch and Herzog in Vienna in 1934. Based on these experimental values, it was then possible to establish a scheme of isotopes of all the elements as they occur in nature.

Rutherford's model of the atom requires that the nucleus carry a positive charge equal to the atomic number of the respective element, which is also equal to the number of electrons surrounding it. As the charge of the nucleus derived from the position of the element in the Periodic Table is always smaller than the atomic weight, the nuclides cannot consist of hydrogen nuclides, that is, of protons alone. There must be particles that either compensate for part of the charge or do not carry any charge at all. Indeed, in 1932, Chadwick in England discovered a particle that had a mass just slightly larger than hydrogen's (i.e., 1.007825 atomic mass units) and no charge at all. This particle, the neutron, has a mass of 1.008665 atomic mass units. It was later found that neutrons are radioactive. A neutron decays with a half-life of 12 min into a proton, an electron, and a neutrino. The neutrino was "invented" to account for energy loss and spin change in β decay.

In general, atomic numbers are denoted by Z, mass numbers and whole numbers denoted by A, and the number of neutrons by N:

$$A - Z = N.$$

Only a limited number of atomic species, of so-called "nuclides", are present in nature; otherwise, there would be an almost infinite number of isotopes for each element. Figure 1 shows which species of each element are present in nature. Thus, our solar system is almost entirely composed of stable species, that is, nuclear species that do not convert

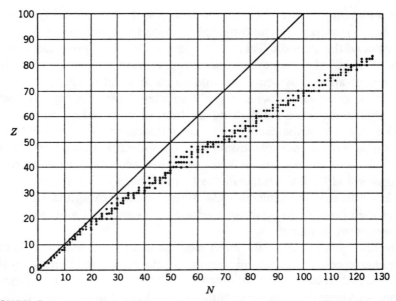

FIGURE 1 Position of the stable nuclear species in an N–Z diagram. The neutron-to-proton ratio increases as a consequence of the coulomb force. All the stable nuclear species occur in nature. (From Friedlander et al., 1966. By permission.)

at ordinary temperatures spontaneously into other kinds. *All stable nuclear species occur in nature.* Also present in nature is a very small fraction of unstable, or radioactive, nuclei, amounting to less than one part per million of the condensable part of solar matter of which the Earth is composed.

The principles determining the stability of nuclear species will be discussed in this chapter. In so doing, and especially in order to explain certain deviations from the general rules, it will be necessary to give an outline of the basic facts of nuclear structure. Much of this information has not yet been reported in textbooks or in review articles for the nonspecialist reader.

1.1.2 Stable and Radioactive Nuclear Species

Since 1932 the assumption that nuclei are composed of neutrons and protons has served as the basis of all theories of nuclear structure. It had been shown by Rutherford that the number of protons in a given nucleus, the *charge number*, or Z, is equal to the atomic number of the particular element in the Periodic Table. With the discovery of the neutron in 1932 it became obvious that the *mass number*, or A, of the nucleus is the sum of the numbers of protons (Z) and neutrons (N):

$$A = N + Z.$$

It is often convenient to use the quantity

$$I = N - Z = A - 2Z,$$

where I is the *neutron excess number*. I is positive or zero for all stable nuclei with the exception of ^{3}He.

Nuclei containing the same number of protons are called *isotopes* and those with the same number of *isotones*. Nuclear species with the same mass number are called *isobars*.

The rules for the existence of stable nuclear species are traditionally expressed in the following form:

1. Rules for the existence of stable isobars
    a. Pairs of stable isobars cannot exist if their charge number differs by only one unit (Mattauch, 1934).
    b. Stable isobars with even mass numbers always have an even number of protons and neutrons (Harkins, 1917).
    c. For each odd mass number, there exists only one stable isobar (Mattauch's rule).
2. Rules for the existence of stable isotopes
    a. For each element, there exists at least one isotope and not more than two isotopes with an odd mass number.
    b. For each element with an even number of protons (even atomic number), there exists also an isotope that contains one more neutron and an isotope that contains one less neutron than the isotope(s) with an odd mass number.
    c. For each element with an even number of protons (even atomic number), there exist also isotopes that contain three (five) more

neutrons and three (five) less neutrons than the isotope(s) with odd mass number (for example, Zn, Ge, W, Hg).

These rules do not constitute entirely independent statements. For instance, Rule 1a follows from 1b and 1c. Rule 2b implies that in each element with even atomic number the even (mass-numbered) isotopes are symmetric with the odd isotope(s) with regard to the sequence of their mass numbers.

There are numerous exceptions to these rules: Rule 1b, for instance, does not hold in the light element region, where ^{6}Li, ^{10}B, and ^{14}N are stable. Contrary to Rule 2a, no odd mass-numbered isotopes exist for four elements (Ar, Tc, Ce, and Pm), two of which (Tc, Pm) possess no stable isotope at all. For many elements the mass numbers of the stable, even isotopes are not symmetric with respect to the one(s) with odd mass number(s), for instance, in samarium, zirconium, and neodymium.

It is easy to understand the rules for the existence of stable nuclear species from the viewpoint of a more general picture of nuclear properties. It will be more difficult to show that it is also possible to understand the many exceptions to these rules.

1.1.3 Binding Energies and Beta Decay

Chemists usually express energy differences associated with chemical reactions in kilocalorie per mole (kcal/mol). The nuclear physicist uses electron volts (eV), defined as the kinetic energy that one electron (or any object carrying one elementary charge unit) obtains when it travels through a potential difference of 1 V. One electron volt is equal to 23 kcal/mol. Hence we may say a chemical reaction is accompanied by an energy change of perhaps 2, 3, or 4 eV. The energies involved in nuclear reactions are about a million times larger. They are given, in general, in MeV (million electron volts). If the reaction energies become that large, then Einstein's mass equivalent becomes important:

$$E = mc^2.$$

The energy of one atomic mass unit ($\frac{1}{12}$ of that of ^{12}C) is *931,4812 MeV*.

One general and fundamental property of nuclei is that a neutron within a nucleus can transmute itself into a proton and conversely a proton into a neutron, if such a transition is energetically possible. The conversion of a neutron into a proton is accompanied by the emission of an electron (a β^- particle) and a neutrino in such a way that the laws of charge, energy, and spin conservation are satisfied. A proton in a nucleus can convert itself into a neutron either by capturing an electron from the atomic K (or L) shell (K-capture process) or by emitting a positron (a β^+ particle). The emission of a β^+ particle is only possible if the energy difference relative to the uncharged system can account for the rest mass of the positron plus that of the excess electron in the resulting negative ion, so this energy difference must be greater than

$$2mc^2 = 1.02 \text{ MeV}.$$

Otherwise, only K capture is possible. In this case, a nucleus "captures" an electron from the K or L shell of the atom, thus decreasing its charge by one unit.

In the following discussion the term *binding energy* of an atom will be used to denote the sum of the binding energies of all the neutrons and protons in the nucleus plus the binding energies of the associated electrons. The total binding energy $E(B)$ of an atom containing N neutrons and Z protons is then

$$E(B) \equiv [NM_n + ZM_p - M(N, Z)]c^2,$$

where M_n, M_p, and $M(N, Z)$ denote the masses of the neutron, the proton (plus one electron), and the atom, respectively. The differences in the binding energies of the electrons in the hydrogen atoms and in the atom (N, Z) can usually be neglected.

The expression β-*decay energy* here denotes the maximum β energy released in any kind of beta transition, including neutrino energies, whereby β^- decay energies are considered positive and β^+ decay energies negative. This β-decay energy E equals the difference of the binding energies of the ground states of two neighboring isobars plus the decay energy of one free neutron ($E_n = 0.2$ MeV):

$$E_\beta = BE(N, Z) - BE(N - 1, Z + 1) + E_n,$$

$$E_\beta = BE(A, I) - BE(A, I - 1) + E_n.$$

1.1.4 The Liquid Drop Model

A semiempirical formula exists for the binding energies of the nuclei that can be derived from a simplified nuclear model, the liquid drop model.* It is

$$BE(A, I) = a_1 A - a_2 A^{2/3} - a_3 I^2/4A - a_4 Z^2/4A^3 + \delta. \qquad (1)$$

In this liquid drop formula the constants have the following meaning:

a_1: heat of condensation = 14 MeV.

a_2: surface tension energy = 13 MeV.

a_3: excess neutron energy = 18.1 MeV.

a_4: coulomb energy = 0.58 MeV.

δ: "even–odd effect" (arising from the existence of "pairing energies" acting between two protons and between two neutrons). For details see Section 1.1.6.

The nuclear stability rules (see Sect. 1.1.2) can be deduced from Equation (1). The δ term in the equation corresponds to the existence of the three different types of nuclear species, as listed in Table II.

In general, the total binding energies $BE(A, I)$ are largest for (e-e) and smallest for (o-o) nuclei. To visualize the binding energies graphically as a function of A and I, or, equivalently, of N and Z, it is necessary to imagine a three-dimensional graph showing the shape of the three energy surfaces as a three-floored U-shaped valley, with the three floors representing the binding energies of the three types of nuclear species. The bottom of this energy valley is in Figure 1 the region populated by stable nuclear species, starting at an angle of 45° and, because of the coulomb energy, slowly bending off with increasing mass numbers into regions where N is increasingly larger than Z.

*See any textbook on nuclear physics or nuclear chemistry–for example, Friedlander et al. (1966).

TABLE II
Number of Stable and Nearly Stable Nuclear Species in Nature

A	Z	N	Symbol	Number of Species in Nature
even	even	even	(e-e)	166
odd	even	odd	(e-o)	53
	odd	even	(o-e)	57
even	odd	odd	(o-o)	7

For the binding energies of isobars (for which A is constant), Equation (1) gives parabolic functions of $N - Z$. In the case of an odd mass number, one particular point for an $N - Z$ value of the one nearest to its bottom will indicate the maximum binding energy, thus representing the highest value of binding energy of any nucleus and therefore the only stable element with this mass number (see the left side of Fig. 2). This explains Rule 1b. In the case of an even mass number, there are two types of nuclei, (o-o) and (e-e), and, correspondingly, two parabolas coexist, differing by 2δ, as shown on the right side of Figure 2. In general, all points on the upper parabola for the (o-o) nuclei, for which N and Z are integers, lie higher than the adjacent points on the lower parabola, so that all the (o-o) nuclei at this mass number have neighboring isobars of the (e-e) type with larger binding energies. Hence all the (o-o) nuclei will be β unstable, as expressed by Rule 1c. This may not be true if δ and the width of the parabola are comparatively small, as in the low-mass-number region. This explains the existence of the stable (o-o) nuclei for $A \leq 14$, as mentioned earlier.

The number of stable isobars at an even mass number depends on the width of the parabola and on the value of the pairing energy at this mass number. In the small-A region, only one isobar is stable. For $A > 46$, two or more stable isobars can exist. Three stable isobars exist at $A = 96$, 130, and 136. Isotope Rules 2a–c follow from this.

The empirical rule for the existence or nonexistence of a nuclear species in nature can be interpreted from the viewpoint of a relatively crude picture of the binding energies derived from the liquid drop model. The stability or instability of a nuclear species is generally determined by its β-decay property. Beta-radioactivity has been found for nearly all nuclear

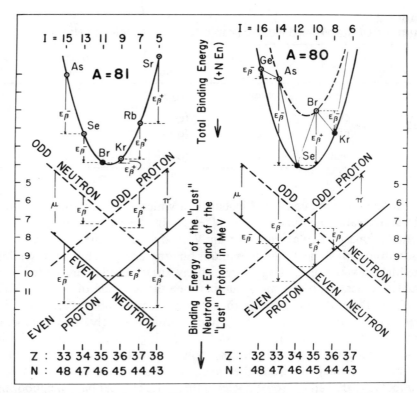

FIGURE 2 Semiempirical graphic illustration of the total binding energies of the isobars of mass number $A = 81$ (*left side*) and 80 (*right side*). The energy values lie on parabolas, a single parabola for odd A and two parabolas for even A. The binding energies of the "last" proton and neutron (including neutron decay energy), respectively, are approximated by straight lines indicated in the lower part of the figure. Pairing energies of protons are denoted by π and of neutrons by μ. The β-decay energies $E_\beta{}^-$ and $E_\beta{}^+$ are indicated.

species with mass numbers ≤ 209 not present in nature. With the exception of the heavy-element region, there is only one case, ^{146}Sm, and possibly a second (^{150}Gd), in which the nonexistence of a nuclear species has to be attributed to α instability rather than β-radioactivity.

The exceptions to these empirical rules show that the liquid drop formula for nuclear binding energies, Equation (1), gives a valuable, if rather crude, approximation to the true situation. The nuclear binding energies are by no means such smooth functions of N and Z as the formula would indicate. If, for example, one attempts to determine the line of maximum stability (the bottom of the Gamow valley) empirically from the positions of the stable odd-A nuclei in an $N - Z$ diagram, then one finds certain regions in which this line winds and twists strangely. In order to look into these questions in greater detail, it is necessary to consider two refinements to the concept of nuclear binding energies: (1) the nuclear shell model and (2) the so-called *pairing effects*, as expressed by the delta term in Equation (1).

1.1.5 Magic Numbers and Nuclear Shell Structure

The empirical evidence for the existence of certain numbers of protons and neutrons, the so-called *magic numbers* that bear a special significance to nuclear structure, can be summarized briefly as follows:

1. The number of stable isotopes and isotones that contain a magic number of either protons or neutrons is exceptionally large (Fig. 1).

2. The abundance of nuclear species containing these magic numbers of protons or neutrons is unusually large (Fig. 14).

3. In the heavy-element region, the significance of magic numbers $Z = 82$ and $N = 126$ can be deduced from α-decay systematics (Perlman et al., 1950).

4. Delayed neutron emission in fission processes occurs in nuclei containing $N^* + 1$ neutrons, where N^* is a magic number of neutrons.

5. Neutron-capture cross sections of nuclei with a magic number of neutrons are exceptionally small (Hughes and Sherman, 1950).

6. The binding energies of the "last" proton or neutron, as derived either by experiment or from β-decay systematics (see below) or by direct measurement, show clear discontinuities in these regions.

7. Practically all nuclear properties change when a magic number is reached. Such properties include spin, magnetic moment, and electrical quadrupole moment, as well as qualities correlated with these properties, such as the occurrence of metastable isomeric states.

Elsasser (1933, 1934) was the first physicist to notice (1) and (2), and he emphasized a possible correlation with nuclear shell structure. Many

Earth scientists were so impressed by the geochemical facts that they made unsuccessful attempts to propose nuclear theories to explain them (Goldschmidt, 1937). In 1948, Maria G. Mayer published convincing evidence for the significance of the magic numbers in nuclear structure, and two years later, she succeeded in postulating a theory to explain them (Mayer, 1950). The same explanation was proposed at exactly the same time completely independently by Haxel, Jensen and Suess (1949, 1950). Its usefulness has since been confirmed by a number of experimental observations. A brief outline of this idea follows.

1. The empirical numbers bearing the magic significance in nuclear structure are 2, 8, 20, 28, 50, 82, and 126.

2. It is easy to see that these numbers belong to two different arithmetical series of N and N^*:

$$N = 2, 8, 20, 40, 70, 112;$$
$$N^* = 2, 6, 14, 28, 50, 82, 126.$$

3. For each series a simple arithmetical law holds, which can easily be recognized by forming the first, second, and third differences:

	2 ... 8 ... 20 ... 40 ... 70 ... 112				
first difference	6	12	20	30	42
second difference	6	8	10	12	
third difference	2	2	2		

	2 ... 6 ... 14 ... 28 ... 50 ... 82 ... 126					
first difference	4	8	14	22	32	44
second difference	4	6	8	10	12	
third difference	2	2	2	2		

Scientists and many laymen are acquainted with Niels Bohr's model of the atom. This model is governed by the idea of attributing to each electron a defined quantum-mechanical state given by the Schrödinger equation. According to the Pauli Principle, each of these states can be occupied by no more than $2j + 1$ electrons, with $j = l \pm \frac{1}{2}$ (where l is the orbital quantum number and j is the total particle spin). The complete filling of all the possible states with a given quantum number is called *shell closure*. The resulting shell structure of the electron envelope of the atom explains the periodicity of the chemical and physical properties of the elements as represented in the Periodic Table.

For a long time, and for various reasons, it seemed impossible that the quantum-mechanical states of individual nucleons in the nucleus could account for shell structures similar to those of the atomic electrons. However, enough evidence has now accumulated to indicate the usefulness of this approach. Just as for the electrons in the coulomb field of an atom, the Pauli Principle holds independently for neutrons and protons in a field of nuclear forces. The quantum-mechanical states of the particles are given by solutions of the Schrödinger equation (see Tables IIIA and IIIB).

The potential, acting upon each particle inside the nucleus, may be approximated by a three-dimensional harmonic oscillator potential. The solutions of the Schrödinger equation may be calculated or can be found in textbooks on quantum mechanics. The shapes of the harmonic oscillator potential, and for comparison, that of the coulomb field in the atom, are shown in Figure 3.

The level sequence of the states of a particle (electron or nucleon) according to the Schrödinger equation is given in Table IIIA for a cou-

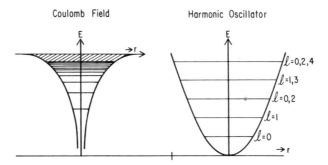

FIGURE 3 Schematic sequence of the quantum-mechanical states of a particle in a coulomb field ($E \propto 1/r^2$) and in a harmonic oscillator field ($E \propto r^2$). See Figure 7 and Tables IIIA and IIIB.

TABLE IIIA
Particles in an Isotropic Coulomb Field

	N=1	N=2			N=3					N=4						
l	0	0	1		0	1		2		0	1		2		3	
j	1/2	1/2	1/2	3/2	1/2	1/2	3/2	3/2	5/2	1/2	1/2	3/2	3/2	5/2	5/2	7/2
State	s^+	s^+	p^-	p^+	s^+	p^-	p^+	d^-	d^+	s^+	p^-	p^+	d^-	d^+	f^-	f^+
No.	2	2	2	4	2	2	4	4	6	2	2	4	4	6	6	8
Σ	2	8			18					32						

l = orbital momentum; $j = l \pm \tfrac{1}{2}$; No. = number of particles in state; Σ = number of particles in shell.

TABLE IIIB
Particles in a Three-Dimensional Harmonic Oscillator Field

N	1	2	2	3	3	3	4	4	4	4	5	5	5	5	5
l	0	1	1	0	2	2	1	1	3	3	0	2	2	4	4
j	1/2	1/2	3/2	1/2	3/2	5/2	1/2	3/2	5/2	7/2	1/2	3/2	5/2	7/2	9/2
State	s^+	p^-	p^+	s^+	d^-	d^+	p^-	p^+	f^-	f^+	s^+	d^-	d^+	g^-	g^+
No.	2	2	4	2	4	6	2	4	6	8	2	4	6	8	10
Σ	2	6		12			20				30				
Total no.	(2)	(8)		(20)			(40)				(70)				

The quantum number l has alternatingly even and odd values as underlined in Part A.
l = orbital momentum; $j = l \pm \frac{1}{2}$; No. = number of particles in state; Σ = number of particles in shell.

lomb field and in Table IIIB for a three-dimensional harmonic oscillator field. According to the Pauli Principle, the number of vacancies for each state is $2(2j + 1)$. The numbers in the last line of Table IIIB give the sum of these vacancies up to the respective level. From this, it would be expected that nuclei containing 2, 8, 20, 40, 70, or 112 protons or neutrons would exhibit shell closures. This expectation must be compared with the empirical evidence.

Our knowledge of the quantum-mechanical state of the electrons in atoms is derived mainly from atomic spectra. The γ spectra emitted by nuclides are much more complex. They cannot be interpreted on the basis of a simple liquid drop model. For a long time, it seemed impossible that a single-particle model, that is, a model of quantum-mechanical states of individual particles, could be useful for deriving meaningful approximations of the properties of nuclear species. It may now be seen that this is not the case. It should be kept in mind, however, that in principle the mixing of two models, the liquid drop model and the shell model, should be avoided, but it is done here as a didactic aid. As illustrated in Figure 4, the binding energies of the "last" neutron for neutron excess numbers 1, 11, and 21 are plotted as a function of the neutron number derived from the liquid drop model.

The shell model explains that, once a magic number is reached with increasing neutron number, an additional neutron exhibits an abrupt drop in its binding energy. In the upper part of Figure 4, the well-known ionization energy of the last electron in the atom is plotted versus its number of electrons. The binding energy of a neutron (and also of a proton) in a nucleus is about a million times larger than that of the last electron in an atom, but the drop in binding energy after a shell closure in the nucleus is, relative to the total binding energy, much smaller than in the atom.

The existence of "shells" in nuclear structure is not considered by many physicists to be as surprising as the fact that this structure can be recognized from the quantum-mechanical state of a single odd particle, despite the fact that in a nucleus the individual nucleons are assumed to interact with all the others. Therefore, the shell model is also called the single-particle model or independent-particle model. Deviations from this model can be recognized in Figure 5. They occur when the states of more than one particle are involved in the overall state of a nucleus. Instead of using the still very incomplete and complex data of nuclear

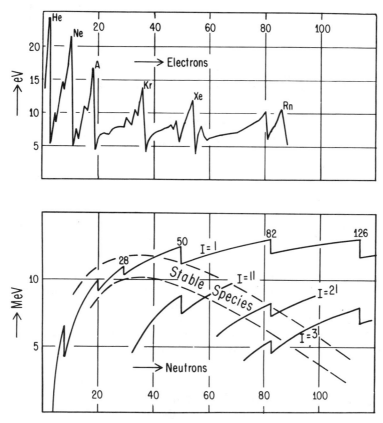

FIGURE 4 *Top*: Ionization energies of atoms drop when electron numbers exceed those of the rare gases. *Bottom*: Drop in the binding energy of the "last" neutron at a shell closure in a nucleus.

spectroscopy, experimental values for spin and parity, primarily for nuclides with odd mass numbers, were used to test the validity and usefulness of a shell model. The spin value of an individual nucleon is $\frac{1}{2}$. The spins of odd-A nuclides can have values of $\frac{1}{2}$, $\frac{3}{2}$, $\frac{5}{2}$, and so on (see Fig. 5). The spin values and magnetic moments of the ground states of even–even A nuclides are always zero; odd–odd A nuclei possess integer spins of 1, 2, 3, and so on. This fact makes it seem plausible that nuclear spins and magnetic moments reflect the quantum-mechanical state of a single odd particle in the nucleus.

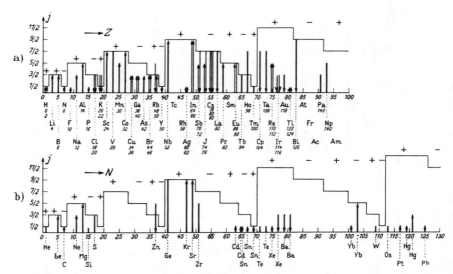

FIGURE 5 Nuclear spin and nucleon number: On the abscissa, numbers of protons (*a*) and neutrons (*b*) are plotted. All the odd-A nucleons with experimentally determined spin values are represented by vertical lines. Their lengths indicate the spin values given on the y-axis. The step-shaped lines indicate the idealized spin values according to the single-particle model. Arrowheads show empirical spin orientations as derived from the Schmidt diagram shown in Figure 6. (Updated from Suess et al., 1949.)

With the assumption that the state of a single particle in an odd-mass-numbered nucleus is responsible for the experimentally observed spin and magnetic moment of the nucleus, this state can then be determined in the following way: In the upper part of Figure 6 the magnetic moments of odd-Z nuclei are plotted against their spin values. This was first done by Schmidt (1937), who found the empirical points in such a plot fell into two groups, one along a higher and one along a lower line, as shown in Figure 6. The two groups represent the two possible orientations of the intrinsic spin of $\frac{1}{2}$ of the odd particle relative to l, its orbital moment. A point close to the upper line indicates that the spin of the nucleon is oriented in such a way that the intrinsic spin is added to that of the orbital moment; a point on the lower line indicates that the spin is subtracted. For example, if the measured spin of an odd nucleus is $\frac{3}{2}$ and its measured magnetic moment is such that the nucleus belongs to the upper group in the upper part of Figure 6, then a $p_{3/2}$ state will have to be attributed to the odd proton. If, however, according to its magnetic moment, the

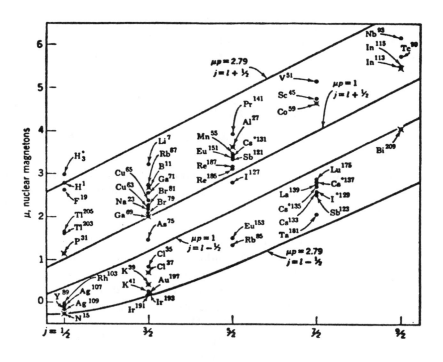

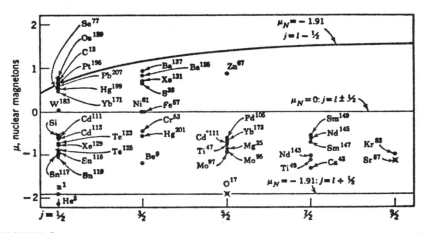

FIGURE 6 Schmidt lines. Magnetic moments of nuclei with odd Z are plotted in the upper part, and of nuclei with odd N in the lower part, against their spin. (After Mayer and Jensen, 1950. By permission.)

nucleus belongs to the lower group of the upper part of Figure 6, then the spectrographic state of the odd proton will be a $d_{3/2}$ state, using the following conventional notation for the orbital moment (the second spectroscopic quantum number):

$$l = \text{orbital moment quantum number} = 0 \quad 1 \quad 2 \quad 3 \quad 4 \quad 5 \quad 6$$
$$\text{notation} : s \quad p \quad d \quad f \quad g \quad h \quad i,$$

In Figure 5, the observed spin values are indicated by vertical lines as a function of the total number of protons in the upper part (a) and of neutrons in the lower part (b). For proton and neutron numbers, vertical lines are shown as arrows that point upwards to indicate parallel orientation and downwards to indicate antiparallel orientation. The broken step-shaped lines indicate the spin values expected from a three-dimensional harmonic oscillator model. The respective spectroscopic states s, p, d, and so on, are denoted by signs (plus for upward, minus for downward). Undoubtedly there is some correlation between observed and expected values, although, just as for electrons in the atoms, the level sequence does not necessarily follow that of the model. Also, in some cases spin and magnetic moment of the ground state of a nucleus result from a combination of the states of three particles rather than from a single one. This seems to be the case for ^{23}Ne and ^{55}Mn. In any case, the single-particle model can now be taken to be a useful starting point for theories of nuclear spectroscopy, just as a liquid drop model is for nuclear energies.

The conception that, in a three-dimensional field of nuclear forces, defined quantum-mechanical states can be attributed to the individual nucleons leads to a straightforward interpretation of the first series of magic numbers: 2, 8, 20, 40, and so on (see Tables IIIA and IIIB). However, this simple model of states of a single particle in a three-dimensional harmonic oscillator field does not lead to the empirical magic numbers for elements heavier than calcium ($A = 40$). Several investigators have therefore attempted to modify the shape of the nuclear field acting upon the individual nucleons in such a way that at least the main magic numbers 50, 80, and 126 can be explained. None of these attempts has been successful. A solution, however, was found by Maria Mayer

(1950) and by Haxel, Jensen, and Suess (1950) on the basis of the empirical evidence described above.

An interpretation is possible by introducing an entirely new principle into the concept of nuclear forces: a strong spin–orbit coupling, the so-called Mayer–Jensen coupling. In the atom, the energy of a state with a given orbital moment is not crucially dependent on the orientation of the spin of the electron relative to this orbit. In the nucleus, however, it must be assumed that the energy state of the nucleon depends strongly on this orientation and that states with parallel spin–orbit orientation are energetically favored (see Fig. 7).

If the states in the harmonic oscillator field are listed in a somewhat different order than given in Table IIIB, so that high spin values are energetically favored, the sequence of the states and the respective particle numbers becomes

1/2	3/2	1/2	5/2	3/2	1/2	7/2	5/2	3/2	1/2	9/2	7/2	5/2	3/2	1/2	11/2
2	4	2	6	4	2	8	6	4	2	10	8	6	4	2	12
(6)		(14)				(28)				(50)					(82)

If the number of particles in the state with the highest spin and a given orbital quantum number l is then added to the number of particles with $l - 1$, the sequence of numbers obtained is exactly that of the second series listed above. Furthermore, it must be assumed that the degree of splitting of the rotational level will depend on the value of l and that this splitting will be greater for large orbital moments than for small ones. Figure 7 illustrates the situation schematically: The left side of the figure shows the term sequence as calculated from a three-dimensional harmonic oscillator model; the center of the figure shows how the terms split according to a strong spin–orbit coupling; and the right side shows the term sequence for a square-well potential . A marked "shell closure" will appear when the energy difference between one level and the next is exceptionally large. Figure 7 shows that these differences are largest when the nucleus contains 2, 8, 20, 28, 50, 82, or 126 particles. A drop in the binding energy occurs when a 3rd, 5th, 21st, 29th, etc., particle

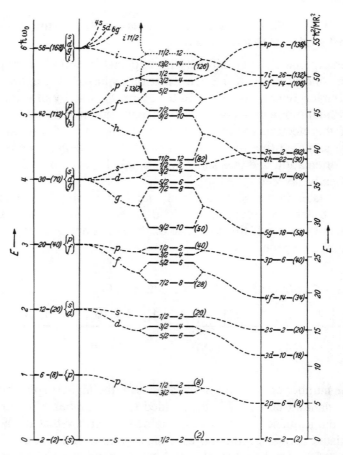

FIGURE 7 Schematic term sequences of a three-dimensional harmonic oscillator (left) and of a square-well potential (right). The fields of the nuclides can be assumed to be intermediate. Splitting of terms as a consequence of a strong spin–orbit coupling increasing with increasing orbital momentum is indicated. Numbers in parentheses give the number of nucleons necessary to fill the shell. (From Haxel et al., 1950.)

is added. The idealized level sequence illustrated in Figure 5 and given above is an extremely simple one and easy to remember. The actual level sequence depends not only on the magnitude, but also on the anharmonicity of the nuclear potential of the spin–orbit splitting, just as for the states of the electrons in the atom. Crossings of levels frequently occur.

The actual spin values for odd-mass-numbered nuclei may deviate considerably from this idealized scheme.

1.1.6 Pairing Effects and the Missing Elements Technetium and Promethium

The liquid drop model leads to completely smooth β-decay systematics, but these do not account for certain irregularities in the energy valley (such as its so-called Gamow windings), nor do they account for the fact that elements 43 (Tc) and 61 (Pm) possess no stable isotopes. The shell model can explain many irregularities but not all of them. No shell closures exist in the regions of proton numbers 43 and 61 and of neutron numbers 19, 35, 39, 45, 61, 89, 115, and 123, for which no stable nuclides exist. A quantitative picture of the irregularities of the energy valley can best be derived from β-decay schemes of unstable nuclear species. They show the differences of the binding energies of the ground states of pairs of isobars. For this difference, the liquid drop formula gives

$$-E_{\beta-} = \mathrm{BE}_{1N}(N, Z) - \mathrm{BE}_{1P}(N - 1, Z + 1) - E_{N}.$$

In this equation, $\mathrm{BE}_{1N}(N, Z)$ is the binding energy of the "last" neutron (in the decaying nucleus) and $\mathrm{BE}_{1P}(N - 1, Z + 1)$ of the "last" proton (in the nucleus formed). E_N denotes the decay energy of the free neutron.

The even–odd effect, expressed by the delta term (δ) in the liquid drop formula, results from "pairing energies", PE, that exist between two nucleons of the same kind, namely, two neutrons or two protons, and also between an unpaired neutron and an unpaired proton. The term δ in Equation (1) will then mean either:

$$\delta = \mathrm{PE}(NN) - \mathrm{PE}(NP)$$

or

$$\delta = \mathrm{PE}(PP) - \mathrm{PE}(NP).$$

Denoting β-decay energies for zero pairing energies by $E_{\beta*}$, then the total β-decay energies for even-A nuclei are

$$E_{\beta-}(\text{o, o}) = E_{\beta*} + \text{PE}(PP) - \text{PE}(NP)$$

$$E_{\beta-}(\text{e, e}) = E_{\beta*} - \text{PE}(NN) + \text{PE}(NP)$$

$$E_{\beta+}(\text{o, o}) = E_{\beta*} - \text{PE}(NN) + \text{PE}(NP)$$

$$E_{\beta+}(\text{e, e}) = E_{\beta*} + \text{PE}(PP) - \text{PE}(NP).$$

Correspondingly, for odd-A nuclei the β-decay energies are

$$E_{\beta-}(\text{odd } N) = E_{\beta*}$$

$$E_{\beta-}(\text{odd } Z) = E_{\beta*} + \text{PE}(PP) - \text{PE}(NN)$$

$$E_{\beta+}(\text{odd } N) = E_{\beta*}$$

$$E_{\beta+}(\text{odd } Z) = E_{\beta*} - \text{PE}(PP) + \text{PE}(NN).$$

The β^- energies are always considered positive and β^+ energies negative.

Figure 2 shows the case of equal pairing energies of the protons and neutrons, $\text{PE}(PP) = \text{PE}(NN)$. This is in general assumed as a first approximation and corresponds to the δ term of Section 1.1.4. With this assumption, only one energy parabola for both kinds of odd-A nuclei is obtained. If, however, the $\text{PE}(PP)$ and $\text{PE}(NN)$ are not equal, then the binding energies of the odd-N and the odd-Z nuclei will not be on the same curve. The energy valley will still be parabolic, but as is always the case for even A, a split into two parabolas will also occur for the odd mass numbers (Fig. 8). The difference of the two parabolas for a given A will be

$$\text{PE}(NN) + \text{PE}(PP), \qquad \text{for odd } A$$

and

$$\text{PE}(NN) + \text{PE}(PP) - \text{PE}(NP), \qquad \text{for even } A.$$

As explained above, a shell closure in nuclear structure results in a sudden drop of the binding energy of the last nucleon, similar to that of the sudden drop in the ionization energy in the case of the atom at a rare gas shell closure (Fig. 4). Hence, whenever a magic number of neutrons or protons is reached, the lines for the binding energies of the last particle

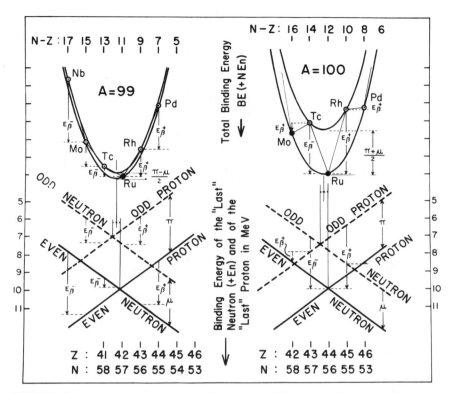

FIGURE 8 Energy parabolas at $A = 99$ and $A = 100$ as examples of a region where neutron and proton pairing energies are *not* equal, so that no stable Tc isotopes exist. See Figure 2 for symbols.

drop abruptly. The situation is presented in Figure 9 for mass number 89, which falls in the region of magic number 50 for neutrons. Pairing energies for the protons and neutrons in this mass range are practically equal, as long as the nuclei contain less than 50 neutrons. In the region $N > 50$, however, PE(PP) is larger than PE(NN) by about 0.6 MeV. Figure 9 shows the resulting shape of the cross section of the energy valley.

In order to obtain a picture of the systematics of decay energies for a range of mass numbers, it is convenient to plot energy differences of ground states of pairs of isobars as derived from β-decay energies versus their mass number. This is done in Figure 10 for odd-mass-numbered

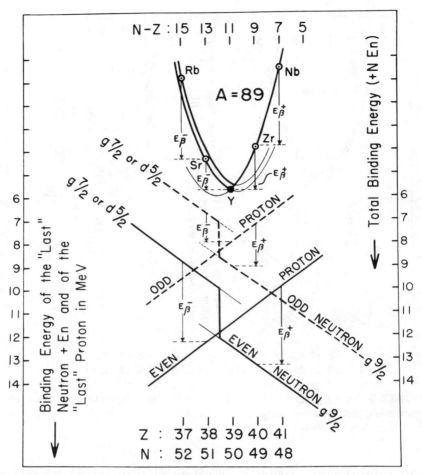

FIGURE 9 Schematic shape of the energy parabolas at the "magic number" region of 50 neutrons (see Figs. 2 and 8 for comparison and explanation).

nuclei for the mass range of $A = 59$ to 159. Squares refer to nuclei with odd neutron numbers and diamonds to those with odd proton numbers. For each point the chemical symbol of the respective unstable isobar is shown. Points referring to nuclear species with the same neutron excess number I are connected, as are nuclides with magic numbers of nucleons. There the lines for constant I show abrupt changes in their trends. Contrary to what is expected from the liquid drop model, these lines also

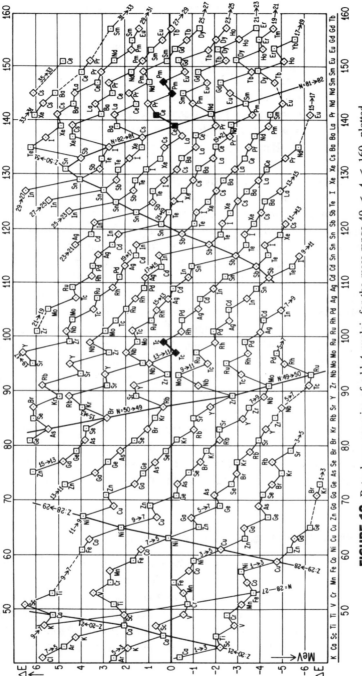

FIGURE 10 Beta-decay energies of odd-A nuclei, for the mass range $40 < A < 160$, plotted versus their mass numbers. Squares denote odd-N and diamonds odd-Z nuclides. Species with the same neutron excess numbers are connected. At the bottom of the figure the chemical symbol of the respective stable isotope is listed for each odd mass number.

show a marked zig-zag pattern in the mass ranges following a magic shell closure. This pattern is caused by the difference in the pairing energies of the two nucleons. In a mass range where proton pairing energies are larger than neutron pairing energies, odd-A species with odd proton numbers will have higher β^--decay energies and lower β^+-decay energies than species with odd numbers of neutrons. If the neutron pairing energies are larger than the proton pairing energies, then the opposite is true.

In general, the following empirical rule can be postulated (Suess and Jensen, 1951): *Whenever the number of one kind of nucleon is somewhat larger than a magic number 28, 50, or 82, the pairing energy of this kind of nucleon will be smaller than that of the other kind.* The neutron excess number I (the isotopic spin) of the stable isobar always changes by two mass units when the lines of Figure 10 cross the zero energy line.

The mass range covered by Figure 10 includes two numbers of protons, namely 43 and 61, that are not present in any stable nuclear species. ^{43}Tc and ^{63}Pm are "missing elements". For a long time, no reason for this peculiarity could be given. Whereas the missing odd-mass-numbered isotope of cerium can be explained easily as a shell effect at $N = 82$, resulting in a β activity of nuclear species containing 79 and 81 neutrons, no shell model could be found to account for the missing elements. The answer can be found in Figure 10, which shows that, as a consequence of a general difference in the pairing energies of protons and neutrons in these particular mass regions, the missing nuclei are accidentally unstable. Just why the pairing energies have this particular character is an interesting question of theoretical nuclear physics that has no simple answer.

1.1.7 Alpha Decay and Fission Processes

Mass numbers and atomic numbers of the nuclides in nature are limited. The largest mass number is that of ^{238}U. The mass of the nuclides is limited by α decay, just as the ratio of neutrons to protons is limited by β decay. The nature of α decay was recognized mainly through the work of Rutherford and Soddy in the early years of this century. Alpha particles are helium nuclei; therefore, the mass number of a nucleus undergoing

α decay decreases by four atomic mass units and its atomic number by two units.

Alpha decay is energetically possible if the binding energy of the "last" two protons and the "last" two neutrons in a nucleus is lower than the total binding energy of all the particles in a helium atom, which is 28.3 MeV. Energetically, this is the case for most nuclear species with $A > 140$. Experimentally, α emission can be observed without difficulty from all nuclei that contain more than 126 neutrons, which is for all nuclei heavier than lead and bismuth. Nuclear species showing α activities and decay energies of the order of 2 MeV and half-lives of 10^{15} years or longer contain 126 or fewer neutrons. Their low activity is difficult to observe. The reason why these nuclei do not decay instantaneously into helium atoms and the residual nuclei whenever this is energetically possible is a potential "wall" formed by an attractive nuclear force potential. The α-decay process is accordingly described by a quantum-mechanical tunneling effect through this potential barrier. The probability of finding a helium atom outside of the wall (outside the range of attractive nuclear forces), and thus the probability of α-decay, was first calculated quantitatively by Gamow (1928). This probability is highly dependent on the energy released by such α-decay processes.

A functional dependence of half-lives on α-decay energies is given by the relationship of Geiger and Nuttall (1911). It holds, fairly accurately, for even-mass-numbered nuclides. The decay of these nuclides almost always leads to the ground state of the daughter. Odd-mass-numbered species usually have a somewhat shorter half-life than even ones. Their decay frequently leads to an excited state of the daughter and is then accompanied by γ radiation from a subsequent transition of the daughter to the ground state.

The α-decay energies as a function of mass numbers show very pronounced magic number effects. Such effects explain the occurrence of a natural α emitter in the rare earth region, as, for example, ^{147}Sm (half-life, 1.06×10^{11} years). Also, the nonexistence in nature of the extinct samarium isotope ^{146}Sm (half-life, 1.03×10^{8} years) is explained by α activity due to the 82 neutron shell closure.

Only three of the nuclear species that contain more than 126 neutrons have survived from the time they were formed by some nuclear processes of element synthesis: ^{238}U (half-life, 4.51×10^{9} years), ^{235}U ($0.713 \times$

10^9 years), and ^{232}Th (13.9×10^9 years). These three long-lived nuclides produce three series of α and β radioactive daughters, ending in stable ^{206}Pb, ^{207}Pb, and ^{208}Pb, respectively. Emission of an α particle decreases the mass number of the decaying nucleus by four. No change of mass number occurs in the case of β decay. Therefore, there must exist four independent α-decay series, two of them consisting of nuclides with even mass numbers, $4n$ and $4n + 2$, and two with odd mass numbers, $4n + 1$ and $4n + 3$, where n denotes a whole number. One of these four series does not occur in nature, as its members had half-lives too short to allow their survival to the present. It is the $4n + 1$ series; ^{244}Pu (half-life, 8.3×10^7 years) is the member in this series with the longest half-life.

The systematics of α-decay of natural and artificial nuclides have been exhaustively investigated by Seaborg and by Perlman and their coworkers at the Radiation Laboratory in Berkeley (Perlman et al., 1950).

Interesting geological features are the "pleochroic halos". These consist of spherical dark areas around radioactive inclusions in certain minerals. They are produced by enhanced darkening at the end of α tracks, where ionization is at its maximum.

The naturally occurring α activities are extremely valuable in geologic research, as they constitute "atomic clocks" for elapsed time measurements of geologic processes.

An unexpected, new mode of nuclear decay was detected in 1938 in experiments in which uranium-containing solutions were irradiated with thermal neutrons. Irradiation of almost all the elements with thermal neutrons produces β activities from one or more isotopes formed by the (n, γ) reactions. Experiments with the uranium compounds, however, had led to a complex mixture of activities believed to come from "transuranium" elements, but this had been impossible to demonstrate. The idea that the energy of about 6 MeV released by the capture of a neutron by the uranium nucleus could induce the breakup of this nucleus had been discussed and also published previously. However, even though the breakup of a uranium nucleus into two fragments must be accompanied by a release of close to 200 MeV, the coulomb barrier to the splitting must be almost equally large. A similar approach to that used for estimations of α-decay half-lives had led to the conclusion that a split into two fragments of similar size should be impossible, even with the help of 6 MeV of neutron capture energy. It is therefore not surprising that this process, so-called nuclear fission, was discovered by chemists and not by physicists, who had discovered the other modes of nuclear decay.

It may also be worth mentioning that the same chemical procedure that had led Madame Curie to the discovery of radium in 1900 also led to the discovery of fission. Madame Curie had found radium in the analytical fraction of barium sulfate and had separated the two elements by fractional crystallization of their chlorides. Hahn and Strassmann suspected the presence of an unknown β-active radium isotope to be present in neutron-irradiated uranium. When, in the fall of 1938, Strassmann tried to separate the radium from the carrier barium, he found that the procedure that had been used by Curie did not work. Finally, after many attempts, Strassmann admitted that he was unable to separate his "radium" from barium by the method that he had used successfully many times before. Was this an unknown effect of some impurity or of something else that changed the property of $RaCl_2$ so that it crystallized with the $BaCl_2$? Strassmann and Hahn continued their attempts to separate the two elements. In December, 1938, they used several other analytical methods, such as bromides and chromates, but found it impossible to separate the radium activity from the barium carrier. Finally, against the advice of many physicists, Hahn and Strassmann published their famous note in January, 1939—and started the Atomic Age.

When uranium or uranium-containing material is irradiated with thermal neutrons, a large number of different radionuclides, the fission products, are formed. By each fission a pair of medium-heavy nuclides form with mass numbers $A(U) = A(F_1) + A(F_2)$ and atomic numbers $Z(U) = Z(F_1) + Z(F_2)$, with F_1 and F_2 denoting the fission products. The ratios of these numbers $A(F_1)/A(F_2)$ and $Z(F_1)/Z(F_2)$, however, vary randomly around a mean value that differs significantly from one. Thermal neutron-induced fission is therefore asymmetric, with the most frequently occurring mass ratio of 1.45. Fission into products with this mass ratio releases a maximum energy of close to 200 MeV. Nearly all fission products form on the neutron-rich side of the energy valley and lead to the β-decay chain. Fission processes yielding a stable nuclide are rare. It was soon discovered that the rare isotope ^{235}U was the one responsible for the fission phenomenon. The much more abundant ^{238}U captures thermal neutrons to form an isotope of element 93: ^{239}Np (half-life, 1.4 days).

The most exciting discovery, however, was that fast neutrons were released during a fission process. The first to observe this was Joliot in Paris, who also immediately recognized the possibility of an atomic chain reaction, at just about the time when World War II started. The intensive work by many of the world's most outstanding physicists that followed was not published until much later. Under the leadership of Enrico Fermi, the first artificial nuclear chain reaction was initiated on December 2, 1942 in Chicago.

Some 30 years later it was discovered that more than a billion years earlier a natural nuclear chain reaction had taken place within a rock formation near Oklo, the Republic of Gabon, West Africa. Kuroda (1982) described this phenomenon in great detail. In most samples, the ^{238}U/^{235}U ratio in terrestrial rocks and minerals is found to be 137.8 ± 0.3 (Senftle et al., 1957). A much higher value for this ratio, 227, was found in the uranium ore from Gabon, and this led to the discovery of this most interesting reaction which took place in Precambrian times.

The capture of one neutron by a ^{235}U nucleus releases enough energy to cause this nucleus to split instantaneously. The capture of a thermal neutron by a ^{238}U nucleus results in its absorption, and no fission occurs unless additional energy is supplied in the form of the kinetic energy of a fast neutron. In general, the smaller the neutron excess number ($N - Z$) of a nucleus, the higher the binding energy of the added neutron and the greater the probability of instantaneous fission.

Is there a finite probability that a uranium nucleus will undergo fission while the nucleus is in its ground state and no energy is added? This question cannot be answered on theoretical grounds, and therefore, as early as 1939, Libby conducted an experiment that gave a lower limit of the partial half-life of 10^{14} years for uranium decay by spontaneous fission. A year later, Flerov and Petrzak (1940) announced the discovery of spontaneous fission of uranium with a partial half-life of 10^{16} to 10^{17} years. It is now known that this partial fission half-life of ^{238}U is 10^{17} years, and that of ^{235}U even longer. Spontaneous fission half-lives decrease rapidly with increasing mass number. Super-heavy nuclear species with charge numbers greater than $Z = 110$ should undergo fission instantaneously. Thus, any buildup process of heavy elements from lighter ones is limited by a recyling mechanism provided by the fission process. No clear indication for such a mechanism can be discerned in the abundance distribution of the medium-heavy elements.

1.2 QUANTITATIVE ANALYSIS

1.2.1 The Isotopic Composition of the Elements

The previous section discussed the kinds of nuclear species that occur in nature. The next question involves their relative natural abundances. We are best informed about abundance ratios of isotopic nuclear species, namely, nuclear species with the same chemical properties. Separations of chemical elements by chemical processes occurring in nature are incomplete, even if the elements have considerably different chemical properties. This explains why traces of all elements may be found almost everywhere. Only the separation of elements from each other into different phases, such as gas and condensates, may occur in a quantitative way. For such cases, the law of the "all-presence" of the chemical elements (I. Noddack, 1936) should not be taken literally. Chemically homologous elements such as zirconium and hafnium, as a rule, are found by geochemical and cosmochemical processes to be only slightly fractionated. The isotopes of a given element, which are essentially identical in their chemical properties, occur in the same abundance ratios in all the materials, terrestrial and extraterrestrial, that have been investigated. The very interesting small deviations that have been observed, constitute an important new field of research that will be discussed in Section 3.1.1. The three rules for the isotopic composition of the elements are generally postulated in the following form:

1. Odd-A isotopes: If an element has two odd-mass-numbered isotopes, the abundances of these two isotopes are nearly equal (Mattauch, 1934).
2. Even-A isotopes: The most abundant even-mass-numbered isotope differs by only one mass number from an odd-A stable isotope.
3. The abundance of an odd-A isotope is always smaller than that of the adjacent A isotope of this element (Harkins, 1917).

In Rule 1, "nearly" means that the abundances differ by not more than a factor of 3. The exceptions are potassium and some elements for which Mattauch's rule for the stability of odd-mass-numbered isobars is violated. It was long suspected that the exceptionally low abundances of ^{115}Sn and ^{187}Os were due to the radioactivity of these nuclides. As Martell and Libby (1950) showed however, [115]Sn and [187]Os are not radioactive, but the other members of the pairs of isobars, i.e., [115]In and [187]Re, are unstable.

The isotopic composition of several elements is graphically shown in Figure 11, in which the logarithm of the percentage of each isotope is plotted against the mass number. For each element the even–even isotopes are connected by a line, as are the two odd-mass-numbered isotopes, when present. The resulting geometric figures give the impression that they are not generated by purely random processes. They seem to show some kind of regularities that, however, cannot be interpreted, just as do the geometric figures of the constellations in the night sky.

A qualitative correlation of binding energies with abundances can easily be discerned. Such a correlation is reflected in the lower abundances of the odd-mass-numbered nuclides compared with their even-mass-numbered neighbors. Also the rule stating that the most abundant isotope of an element is, in most cases, the one nearest to the bottom of the energy valley, indicating a correlation with binding energies. However, attempts to derive quantitative relationships have failed (see Sect. 2.1). Other indications of hidden regularities can be appreciated in the following way: The geometry of the figures such as shown in Figure 11 is fre-

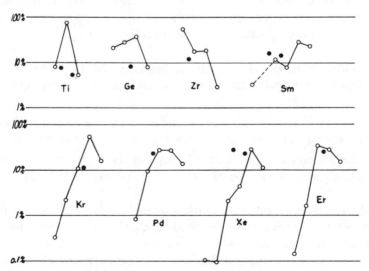

FIGURE 11 Isotopic composition of some elements shown graphically.

quently "similar" for neighboring even-Z elements. The character of the figures changes only gradually with their atomic numbers, with the change being more pronounced in magic number regions. This can be recognized best in Figure 16. In order to obtain further indications of intrinsic laws governing the isotopic composition of the elements, one might plot the change in the abundance of a nuclear species when two neutrons are added against the number of neutrons present in the nucleus to which the two neutrons are added, as has been done in Figure 12. The total number of neutrons was taken as one coordinate, and the effect on the abundance caused by the addition of two neutrons as the other. The points shown are derived from mass spectrometric measurements and are therefore extremely accurate. The connecting lines are drawn between points referring to pairs of nuclei with the same neutron excess.

Figure 12, like Figure 11, visualizes universal quantities, and, although no simple rules are apparent from this graph, it may be conjectured that inherent laws govern this type of distribution of abundance ratios. Clearly, the situation is complicated, yet it undoubtedly demonstrates that regularities do exist in the isotopic composition of the elements. These regularities have nothing to do with chemical properties. They must pertain to nuclear processes that led to the formation of the nuclear species in their present proportions. We therefore expect the same, or at least similar, regularities to govern the relative abundances of the individual nuclear species and thus the relative abundances of the elements.

1.2.2 Empirical Values for the Relative Abundances of the Elements

From the well-known isotopic composition of the elements and from their less well-known relative amounts in our solar system, we can easily calculate how much of each nuclear species is present. Several kinds of observations provide evidence about the relative abundances of each elements. The first, most direct way is the spectral analysis of sunlight. The second is chemical analysis of material that we can lay our hands on terrestrial and lunar rocks and meteorites. Further, less direct ways involve data on the densities and angular momenta of planets and their satellites, as well as data on the composition of solar wind and cosmic rays.

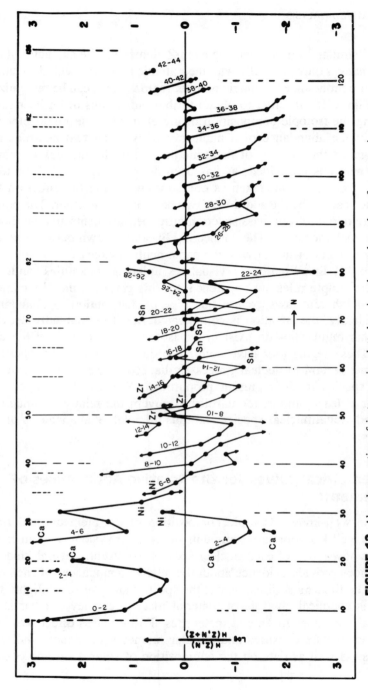

FIGURE 12 Abundance ratios of isotope pairs plotted against the total number of neutrons in the lighter member of the pair (see text for discussion).

The importance of nuclear abundance values becomes obvious when we consider that the near-constancy of the isotopic composition of the elements means that everything we have physically available, and presumably all the matter in our solar system, must have come essentially from material that at one time in the distant past was chemically homogeneous. The bulk of the matter from which the Sun and the other members of the solar system formed must have been essentially homogeneous at one time, even though this matter, as we believe now, is composed of several components that have different nuclear histories.

1.2.3 Astronomical Spectral Analyses

A direct way to determine the relative abundances of the elements in the Sun and stars is by quantitative spectral analysis using absorption lines (Frauenhofer lines). The intensities of these lines are a function of the concentration of the atoms causing the absorption. In general, it is not difficult to identify the element and its spectroscopic state that cause an observed absorption line. The intensity of the line is a complicated function of both the physical properties of the atom and the thermodynamic state of the absorbing matter (Unsöld, 1977). The most important atomic property is the matrix element for the spectroscopic transition under consideration, that is, the transition probability or oscillator strength, commonly called the f-value. This f-value can be derived from direct measurements in the laboratory, or it can be calculated from quantum mechanics to within a certain limit.

The accuracy of the determination of the relative abundance of an element by spectral analysis depends crucially on the accuracy of the f-value of the respective element, used for the determination. The f-values are obtained in the laboratory by measuring the light absorbed by an atomic beam perpendicular to the light beam. In this way Doppler effect and line broadening due to collision damping can be estimated, and corrections can be applied. The "curve of growth", that is the line intensity as a function of absorbing atoms per square centimeter, can be derived by estimating mean values of temperature, electron pressure, and effective thickness of a model atmosphere that will be different for different stellar objects.

But even with an accurate knowledge of the physical properties of the atom, the relative amounts of the atoms in the stellar atmosphere cannot be deduced in a simple way from the relative intensities of the spectral lines. Accurate knowledge of the temperature and pressure in the absorbing layer is necessary, and in order to find these values, some knowledge of the chemical composition of the medium is required. This complex situation can be mastered only by long and complicated stepwise deductions and successive approximations. Such quantitative spectral analyses of the light of the stars and the Sun were first applied by Menzel (1931), Strömgren (1940), and Unsöld (1944) in order to improve the classical work of Russell (1929) and Payn (1925) in this field. Later, a number of investigators (see Aller, 1961) analyzed many more stars. The ultimate aim of this work was to see whether the chemical composition is uniform throughout the universe or whether there are distinct variations. We now know that the chemical compositions of stars and galaxies do vary. The astrophysical interpretation of these variations is at present an important field of research (see Sec. 3.1.1).

The abundance values derived from the absorption lines in the solar spectrum are not as accurate as those from chemical analyses of rocks and meteorites. Nearly all elements are present in these solids. Yet the most abundant elements of all, hydrogen and helium, are highly depleted in meteorites and terrestrial matter. Also, carbon, nitrogen, oxygen, and neon are much less abundant than in the Sun and the universe relative to the heavier elements (called metals by astronomers). Together with hydrogen and helium, they constitute more than 99% of the mass of the solar system. We know their abundance values only from astronomical observations of the absorption lines in the spectrum of the Sun. Our knowledge of the abundances of the most abundant "atmophile" elements—hydrogen, helium, carbon, nitrogen, oxygen, and neon—rests entirely on astronomical observations. For some time the abundance of iron in the Sun seemed to be a mystery, as the solar values were nearly four times smaller than those in meteorites and terrestrial planets, as derived from their densities (see Sect. 4.3.3). As Urey (1952) discussed in great detail, this large difference, if real, required certain assumptions for the separation of metal from silicate, which appeared unlikely to many investigators. Much to their relief, however, it was discovered that, as a result of a trivial experimental mistake, the f-value for iron that had been used to calculate the solar iron abundance was incorrect. The small furnace used to produce the beam of iron atoms had a temperature higher by several degrees than had appeared from the measurements. The iron abundance in the Sun using the corrected f-value now corresponds to that observed for the other members of the solar system.

1.2.4 Empirical Nuclear Abundance Rules Derived from Trace Element Concentrations in Meteorites

As noted earlier, the abundances of the elements, and more conspicuously the abundance ratios of even-mass-numbered isotopic nuclear species, show unmistakable signs of some type of mediation with nuclear properties (see Fig. 12). In particular, these ratios show dramatic effects at the magic numbers of 28, 50, and 82 neutrons. It is to be expected that the abundances not only of isotopic species, but also of all nuclear species show regularities of this kind. Approximate empirical values for the relative abundances of all nuclear species can be obtained from the elemental abundance ratios and from the isotopic compositions of the elements. When these nuclear abundance values are plotted as a function of mass number using Goldschmidt's elemental abundance values (Goldschmidt, 1937), a somewhat confusing picture emerges; this difficulty, however, can be greatly simplified by using only the odd-mass-numbered nuclear species. At each odd mass number, only one isobar is stable, and, therefore, only one abundance value exists. Part a of Figure 13 shows the log of these odd-A abundance values plotted against their mass number A.

Immediately, one important feature can be recognized: a smooth dependence of abundances on mass number in the mass region of $A = 140$–178. This region is that of the odd-A rare earth isotopes, namely, isotopes of elements that have very similar chemical compositions.

Chemists find it difficult to separate rare earth elements quantitatively from each other in the laboratory. Therefore, it is not surprising that in nature these elements frequently occur in nearly constant abundance ratios. In most meteorites, it appears that rare earth elements have not been fractionated appreciably from each other by cosmochemical processes. Hence, their abundance ratios closely represent the primordial ratios in which these elements formed during nuclear synthesis. The smooth dependence of abundances of the odd-mass-numbered isotopes therefore corresponds to the yields of the nuclear processes that led to their formation.

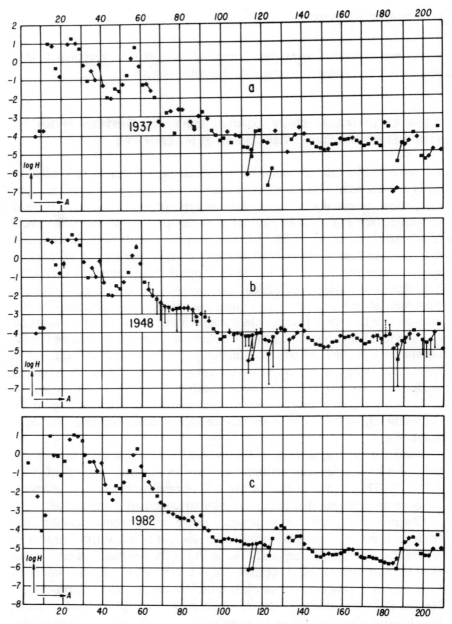

FIGURE 13 The $_{10}$log of the abundances of the odd-A isotopes relative to H(Si) = 100 are plotted versus their mass numbers. a: Data were derived from abundance values given by Goldschmidt (1937). b: Values were modified according to Rule 1 (Suess, 1947). c: Values from chemical analyses of carbonaceous chondrites (Palme et al., 1981.).

Thus, the suspicion appeared justified that this smooth dependence of abundances with mass numbers should prevail for all the odd-mass-numbered species of the Periodic Table. Deviations from such regularity should then be due to cosmochemical fractionation processes. With this idea in mind, Suess (1947) attempted to estimate the magnitude of the effects of cosmochemical fractionation processes (and also experimental analytical errors in trace element determinations). He suggested a dependence of the actual primeval abundance on mass number, as shown in Figure 13, part b. This, indeed, was most exciting, as it made it possible to recognize other regularities. It was found that if abundance values for odd-A elements were chosen in such a way that their odd-mass-numbered isotopes showed a smooth behavior, then at the same time the even-mass-numbered isotopes also exhibited similar regularities in their abundances, though not as conspicuous as those of the odd-A species. Several years later Suess and Urey (1956), guided by the idea that all nuclear abundances should vary in some regular way, revised the analytical data for meteorites in such a way that an abundance picture was obtained that showed regularities and that could still be considered within the limits of error of the analytical data. The abundance rules, which had originally been published in German by Suess (1947), were formulated in the following way:

1. *Odd-mass-numbered nuclides.* The abundances of odd-mass-numbered nuclear species with $A > 50$ change steadily with the mass number. When isobars occur, the sum of the abundances of the isobars must be used instead of the individual abundances.

2. *Even-mass-numbered nuclides.* In the region of the heavier elements with $A > 90$, the sums of the abundances of the isobars with even mass numbers change steadily with mass number.

3. In the region of the *lighter elements* with $A < 70$, the isobar with the higher excess of neutrons is the less abundant one at each mass number. In the region of the heavier elements with $A > 70$, the isobar with the smallest excess of neutrons is the least abundant.

4. *Exceptions to these rules* occur at mass numbers where the numbers of neutrons have certain values, the so-called magic numbers.

All subsequent work dealing with the relative abundance of the elements has used these rules as a guide for estimating the original primordial abundance values. The first rule appeared to be especially useful for this purpose. It was most impressive to see how the analytically determined values for certain types of meteorites—in particular, type I carbonaceous chondrites—approached with increasing experimental accuracy results that agreed with these rules. Recent measurements of the trace element content of carbonaceous chondrites correspond so closely to these rules as to definitely rule out the possibility that cosmochemical processes have changed the primordial amounts of these elements significantly.

With this firm knowledge of primordial nuclear abundances, it should now be possible to explain the abundance rules from the viewpoint of theories of elemental synthesis. Deviation from these rules may indicate that the need for refinement of the theories. In Figure 13, part C, the now experimentally determined abundance values for the odd-mass numbered species are plotted. These values are derived from trace element analyses of carbonaceous type I chondrites, primarily from the meteorite Orgueil that fell in France in 1864. Figure 14 presents a rough overview of the general abundance distribution of the elements as derived from Goldschmidt's original data. The upper part shows a plot for each mass number of the sum of isobaric abundance values. These values are Goldschmidt's modified with the assumption that the abundance rules are valid.

For the light and most abundant volatile elements—oxygen carbon, nitrogen, and argon—astronomical data have been used. The lower part of Figure 14 shows the $\log_{10}$ of abundances versus mass numbers, as if N versus Z, the atomic number, were plotted, except that here the coordinates are tilted 45°. In this way, isobars appear on the same X-coordinate. Just as if we were to consider the binding energies plotted on a third coordinate, a Z-coordinate, and visualize an energy surface in the form of a steep valley (see Sec. 1.1.4), we can imagine the logarithm of the abundances plotted on a third, a Z-axis. The shape of the abundance surface can then be imagined as having the following form: The abundances of the odd-mass-numbered species are represented by a single value per mass number. They form a more-or-less smooth line. The distribution of the abundance values of the even species is presented in the form of a topographic map whose lines indicate constant abundance. The graph indicates that the surface of the abundance values is much less regular than the energy surface. The principal features of this abundance surface are the following: The values for hydrogen and helium are

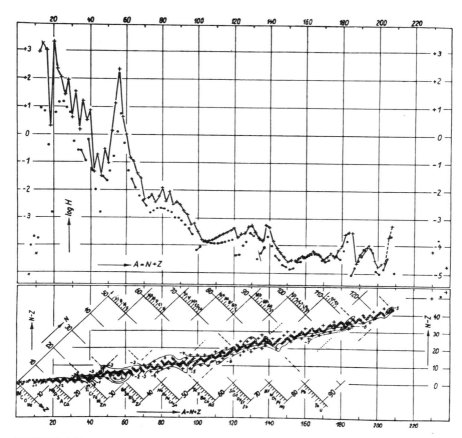

FIGURE 14 *Top*: Logarithms of abundance values of the individual nuclear species plotted for even-*A* species (+) and for odd-*A* species (•) versus the mass number *A*. *Bottom*: Positions of stable (and nearly stable) nuclides are indicated in an *A/N-Z* diagram, equivalent to an *N/Z* diagram tilted 45°. The abundances of the individual nuclides are qualitatively illustrated by a topographic map with the logarithms of the abundances as a third coordinate. Lines of equal abundances are drawn for each power of ten. (From Suess, 1949a).

off scale at the upper-left corner; the atomic abundance of hydrogren is higher than that of any other element (12.5 on this logarithmic scale). The isotopes of boron, lithium, and beryllium are more than 10 orders of magnitude less abundant. These and a few other nuclides are the ones that participate in thermonuclear reactions in the interior of ordinary stars.

The diagram is meaningful only for mass ranges where more than one stable isobar exists; otherwise, there is only a line from the hydrogen corner downwards 4 or 5 orders of magnitude to the steep downgrade in the direction of increasing A. There is an abrupt decrease in abundances beyond ^{40}Ca leading to a deep valley between mass numbers 40 and 50. Thereafter, with increasing mass number, abundances rise to a single peak at mass number 56—one high needle, like the top of a steep mountain, occupied by ^{56}Fe. From this peak, abundance values drop rather smoothly with increasing mass number until $A = 70$ is reached, namely, the gallium–germanium region of the Periodic Table. The next remarkable feature of this topographic map is a ridge around mass number 190. The significance of this ridge is that the nuclides that make it up all contain 50 neutrons. Another ridge is presented by the tin isotopes that contain 50 protons. The most important feature in connection with the theory of formation of the elements are these broad humps of the abundances in the region of $A = 130$ and $A = 195$. Clearly, all these features have to do with magic numbers. This is discussed here in order to show that these features can already be recognized from the relatively inaccurate elemental abundance data given by Goldschmidt (1937).

TWO

Element Synthesis

2.1 EARLY THEORIES

What were the processes that led to the formation of the matter that surrounds us, the chemical elements, and their isotopes of which our astronomical environment consists? Two avenues are usually pursued in approaching this question:

1. One may ask: What, within the framework of physics and given the structure of the expanding universe and the constitution of stars, could have led to the formation of matter as we know it? This has typically been the way astrophysicists and astronomers have approached the question of the origin of the elements.

2. Alternatively, one can start with the chemical nature of the matter surrounding us and then consider how the individual chemical elements are isotopically composed. From these data the relative abundances of the individual nuclear species can be derived. As was shown previously, these abundances are now fairly well known. On the basis of this knowledge, we may then proceed to ask: What are the processes that may have led to the prevailing mixture of nuclear species? Where can these processes have occurred in the universe? And how long ago might this have happened?

There are now some 200 numerical values known to an accuracy of a few per mil for the abundance ratios of pairs of isotopes of elements; while the accuracies of the values of the abundance ratios of the 82 stable chemical elements in the Periodic Table are lower, they are still within reasonable limits of error of 2 to 200%. Table IV lists estimates for "cosmic", or better, "solar system" abundances published between 1937 and 1961. The most recently published data are given in Table V. The differences between the two new series rarely exceeds a factor of two. These numerical values must give us some information about the kinds of nuclear processes that led to the formation of the individual nuclear species in their present relative amounts. The question is, What do these data tell us? What can we possibly learn from them? We shall see that it is possible to get firm answers to some of these questions, but only tentative conjectures in response to others.

As chemists, we first ask: Can we discern from these rigorous numerical data certain types of nuclear reactions that must have been important in the formation of solar system matter? Are there indications that at one time prevailing conditions lead to an approach to thermodynamic equilibrium of the different nuclear species? Do the nuclear abundances reflect "frozen-in" equilibrium concentrations? Attempts to answer these questions have been made by Urey and Bradley in 1931, by Jensen and Suess in 1944, and by Ubbelohde in 1948. Following a suggestion by von Weizsäker, these investigators considered neutron exchange equilibria of the following type, in which different isotopes of the same element appear on each side of the equation:

$$^{17}O + {}^{12}C = {}^{16}O + {}^{13}C.$$

The equilibrium constant K is

$$K = ([{}^{16}O][{}^{13}C])/([{}^{17}O][{}^{12}C]),$$

$$K = g \exp{(-\Delta E/RT)},$$

where g is equal to $\Delta S/R$ with R the gas constant and ΔS the entropy change. The relative amounts of these nuclear species, denoted here as $[{}^{17}O]$, $[{}^{12}C]$, $[{}^{16}O]$, and $[{}^{13}C]$ and the binding energies, E, of the last neutrons are known, which gives the heat of reaction, ΔE.

TABLE IV

Early Estimates of Abundances of the Elements Based on Meteorite Analyses and Astronomical Observations[a]

	Goldschmidt (1937)	Brown (1949)	Urey (revised) (1952)	Aller (astronomical) (1961)
1 H		3.5×10^{10}		2.94×10^{10}
2 He		3.5×10^{9}		4.05×10^{9}
3 Li	100		100	0.6
4 Be	20		16	1.0
5 B	24		20	1580
6 C		8.0×10^{6}		2.7×10^{6}
7 N		1.6×10^{7}		4.9×10^{6}
8 O		2.2×10^{7}		1.58×10^{7}
9 F	1500	9000	300	
10 Ne		2.4×10^{7}		1.73×10^{7}
11 Na	4.42×10^{4}	4.62×10^{4}	4.38×10^{4}	7.7×10^{4}
12 Mg	8.7×10^{5}	8.87×10^{5}	9.12×10^{5}	1.78×10^{6}
13 Al	8.8×10^{4}	8.82×10^{4}	9.48×10^{4}	7.4×10^{4}
14 Si	1.0×10^{6}	1.0×10^{6}	1.0×10^{6}	1.0×10^{6}
15 P	5.8×10^{3}	1.3×10^{4}	5.0×10^{3}	1.9×10^{4}
16 S	1.14×10^{5}	3.5×10^{5}	9.8×10^{4}	5.2×10^{5}
17 Cl	4000–6000	17000	2100	300000
18 A		2.2×10^{5}		1.0×10^{5}
19 K	6900	6930	3160	3900
20 Ca	5.71×10^{4}	6.7×10^{4}	4.90×10^{4}	8.3×10^{4}
21 Sc	15	18	28	42
22 Ti	4700	2600	2440	1800
23 V	130	250	220	300

TABLE IV
Continued

	Goldschmidt (1937)	Brown (1949)	Urey (revised) (1952)	Aller (astronomical) (1961)
24 Cr	1.13×10^4	9.5×10^3	7800	1.9×10^3
25 Mn	6600	7700	6850	5600
26 Fe	8.9×10^5	1.83×10^6	6.00×10^5	4.8×10^5
27 Co	3500	9900	1800	2200
28 Ni	4.6×10^4	1.34×10^5	2.74×10^4	4.4×10^4
29 Cu	460	460	212	932
30 Zn	360	160	180	2880
31 Ga	19	65	11.4	2.5
32 Ge	190	250	65	25
33 As	18	480	4.0	
34 Se	15	25	24	
35 Br	43	42	49?	
36 Kr				1
37 Rb	6.8	7.1	6.5	
38 Sr	40	41	18.9	
39 Y	9.7	10	8.9	
40 Zr	140	150	54.5	
41 Nb	6.9	0.9	0.8	
42 Mo	9.5	19	2.42	
44 Ru	3.6	9.3	2.1	

Element			
45 Rh	1.3	3.5	0.71
46 Pd	1.8	3.2	1.3
47 Ag	3.2	2.7	0.35
48 Cd	2.6	2.6	1.9
49 In	0.23	1.0	0.26
50 Sn	29	62	1.33
51 Sb	0.72	1.7	0.12
52 Te	0.2		0.16
53 I	1.4	1.8	1.5
54 Xe			
55 Cs	0.1	0.1	1.3
56 Ba	8.3	3.9	8.8
57 La	2.1	2.1	2.1
58 Ce	5.2	2.3	2.3
59 Pr	0.96	0.96	0.96
60 Nd	3.3	3.3	3.3
62 Sm	1.15	1.2	1.1
63 Eu	0.28	0.28	0.28
64 Gd	1.65	1.7	1.6
65 Tb	0.52	0.52	0.52
66 Dy	2.0	2.0	2.0
67 Ho	0.57	0.57	0.57
68 Er	1.6	1.6	1.6
69 Tm	0.29	0.29	0.29
70 Yb	1.5	1.5	1.5
71 Lu	0.48	0.48	0.48

TABLE IV
Continued

	Goldschmidt (1937)	Brown (1949)	Urey (revised) (1952)	Aller (astronomical) (1961)
72 Hf	1.5	0.7	0.55	
73 Ta	0.40	0.31	0.32	
74 W	14.5	17.0	13.0	
75 Re	0.12	0.41	0.05	
76 Os	1.7	3.5	0.97	
77 Ir	0.58	1.4	0.31	
78 Pt	2.9	8.7	1.5	
79 Au	0.27	0.82	0.140	
80 Hg	0.33		<0.006	
81 Tl	0.17		0.11	
82 Pb	9.1	<2.0	0.47	
83 Bi	0.11	0.21	0.144	
90 Th	0.59			
92 U	0.23	0.02		

[a][Si $\equiv 10^6$].
Adapted from Suess and Urey (1956).

TABLE V

Primordial Solar System Abundances of the Elements

Z	Element	Normalized Abundance	
		Palme et al. (1981)[a]	Anders and Ebihara (1982)[a]
1	H	2.5×10^{10}	2.72×10^{10}
2	He	2.0×10^{9}	2.18×10^{9}
3	Li	5.5×10	5.97×10
4	Be	7.3×10^{-1}	7.8×10^{-1}
5	B	6.6	2.4×10
6	C	7.9×10^{6}	1.21×10^{7}
7	N	2.1×10^{6}	2.48×10^{6}
8	O	1.7×10^{7}	2.01×10^{7}
9	F	7.1×10^{2}	8.43×10^{2}
10	Ne	1.4×10^{6}	3.76×10^{6}
11	Na	5.7×10^{4}	5.70×10^{4}
12	Mg	1.01×10^{6}	1.075×10^{6}
13	Al	8.0×10^{4}	8.49×10^{4}
14	*Si*	1.00×10^{6}	1.00×10^{6}
15	P	8.6×10^{3}	1.04×10^{4}
16	S	4.8×10^{5}	5.15×10^{5}
17	Cl	5.0×10^{3}	5.24×10^{3}
18	Ar	2.2×10^{5}	1.04×10^{5}
19	K	3.5×10^{3}	3.77×10^{3}
20	Ca	5.9×10^{4}	6.11×10^{4}
21	Sc	3.5×10	3.38×10
22	Ti	2.4×10^{3}	2.40×10^{3}
23	V	2.9×10^{2}	2.95×10^{2}
24	Cr	1.35×10^{4}	1.34×10^{4}
25	Mn	8.7×10^{3}	9.51×10^{3}
26	Fe	8.6×10^{5}	9.00×10^{5}
27	Co	2.2×10^{3}	2.25×10^{3}
28	Ni	4.8×10^{4}	4.93×10^{4}
29	Cu	4.5×10^{2}	5.14×10^{2}
30	Zn	1.40×10^{3}	1.26×10^{3}
31	Ga	4.4×10	3.78×10
32	Ge	1.13×10^{2}	1.18×10^{2}
33	As	6.5	6.79
34	Se	6.3×10	6.21×10
35	Br	8.0	1.18×10
36	Kr	2.5×10	4.53×10
37	Rb	6.4	7.09

TABLE V
Continued

Z	Element	Normalized Abundance	
		Palme et al. (1981)[a]	Anders and Ebihara (1982)[a]
38	Sr	2.6×10	2.38×10
39	Y	5.4	4.64
40	Zr	1.10×10	1.07×10
41	Nb	8.5×10^{-1}	7.1×10^{-1}
42	Mo	2.5	2.52
44	Ru	1.8	1.86
45	Rh	3.3×10^{-1}	3.44×10^{-1}
46	Pd	1.32	1.39
47	Ag	5.0×10^{-1}	5.29×10^{-1}
48	Cd	1.30	1.69
49	In	1.74×10^{-1}	1.84×10^{-1}
50	Sn	2.4	3.82
51	Sb	2.7×10^{-1}	3.52×10^{-1}
52	Te	4.8	4.91
53	I	1.16	9.0×10^{-1}
54	Xe	6.1	4.35
55	Cs	3.7×10^{-1}	3.72×10^{-1}
56	Ba	4.2	4.36
57	La	4.6×10^{-1}	4.48×10^{-1}
58	Ce	1.20	1.16
59	Pr	1.8×10^{-1}	1.74×10^{-1}
60	Nd	8.7×10^{-1}	8.36×10^{-1}
62	Sm	2.7×10^{-1}	2.61×10^{-1}
63	Eu	1.00×10^{-1}	9.72×10^{-2}
64	Gd	3.4×10^{-1}	3.31×10^{-1}
65	Tb	6.1×10^{-2}	5.89×10^{-2}
66	Dy	4.1×10^{-1}	3.98×10^{-1}
67	Ho	9.1×10^{-2}	8.75×10^{-2}
68	Er	2.6×10^{-1}	2.53×10^{-1}
69	Tm	4.1×10^{-2}	3.86×10^{-2}
70	Yb	2.5×10^{-1}	2.43×10^{-1}
71	Lu	3.8×10^{-2}	3.69×10^{-2}
72	Hf	1.8×10^{-1}	1.76×10^{-1}
73	Ta	2.1×10^{-2}	2.26×10^{-2}
74	W	1.27×10^{-1}	1.37×10^{-1}
75	Re	5.2×10^{-2}	5.07×10^{-2}
76	Os	7.2×10^{-1}	7.17×10^{-1}

TABLE V
Continued

		Normalized Abundance	
Z	Element	Palme et al. (1981)[a]	Anders and Ebihara (1982)[a]
77	Ir	6.5×10^{-1}	6.60×10^{-1}
78	Pt	1.42	1.37
79	Au	1.9×10^{-1}	1.86×10^{-1}
80	Hg	4.0×10^{-1}	5.2×10^{-1}
81	Tl	1.7×10^{-1}	1.84×10^{-1}
82	Pb	3.1	3.15
83	Bi	1.36×10^{-1}	1.44×10^{-1}
90	Th	3.2×10^{-2}	3.35×10^{-2}
92	U	9.1×10^{-3}	9.0×10^{-3}

Values are derived from solar spectral analyses (see Sect. 1.2.3) and chemical analyses of the type 1 carbonaceous chondrite Orgueil (see also Fig. 16).
[a] Reprinted with permission.

To calculate the temperature at which the equilibrium was established, a value of g has to be estimated by comparing a number of different neutron exchange reactions. Figure 15 shows values of ΔE plotted against the logarithms of the ratios of abundances of one isotope to the next heavier one. The slope of a line drawn through the points would then indicate a kT of about 0.5 MeV (i.e., $T \sim 5 \times 10^9$ °K), but g contains entropies. Hence, the partition functions for the individual nuclides are

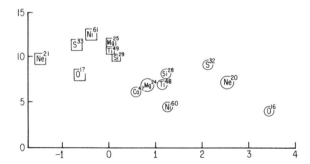

FIGURE 15 Decadic logarithm of the abundance ratios of nuclei to their respective isotopes containing one more neutron plotted against the binding energy (in MeV) of those neutrons in the heavier isotopes.

difficult to estimate. The points in Figure 15 may well be affected by differences in g, so it is doubtful that the temperature obtained from Figure 15 in this way is physically meaningful.

In any case, it is now clear that there are no pressure and temperature conditions that lead to a reasonable approximation of the total nuclear abundance distribution that could have resulted from a frozen-in thermodynamic equilibrium.

The next step in this approach was to investigate the possibility that the kinetics of nuclear reactions could have led to the empirically determined abundance distribution. Obviously what was required was a scenario that explained, at least partially, the observed features as expressed by the abundance rules (see Sect. 1.2.4). For the purpose of interpreting these rules, it is convenient to classify the stable nuclear species as done earlier into:

1. even-A species
2. odd-A species.

The even-A species can be divided into those "shielded" by an isobar in such a way that no additional amounts of the species can be added through β decay. These amounts are retained by the isobars "shielding" them. The shielded species cannot form from unstable isobars with higher neutron excess numbers (for example, fission products). The isobar with the lower neutron excess is shielded by the isobar of greater neutron excess. The shielding even-A species, of course, can form from any parent along the β-decay chain. For example, at $A = 110$, ^{110}Pd is shielding ^{110}Cd against β-decay, so that ^{110}Cd has lower abundance than it might otherwise have, whereas all the other cadmium isotopes can be reached by series of β decays of fission products (see Sect. 1.1.7 for a discussion of fission). It can easily be recognized that, in the medium-heavy- and heavy-element regions, odd-A isotopes, as a rule, are less abundant than even-A ones (Harkins rule) and also that the shielding even-A species are generally more abundant than their shielded isobars. Obviously, a large part of shielding isobars had formed on the neutron-rich, β-unstable side of the energy valley. Attempts to explain these values by one specific mechanisms have failed, even with parameters changing with time. A neutron flux that would account for the higher abundance of the shielding even-A species would inevitably "eat up" most of the nuclides with an odd mass number. Also, many features, such as the high abundances of species with a magic number of neutrons, would inevitably be erased by subsequent high concentrations of free neutrons. It seemed that there was no possibility of explaining the abundance distribution of the nuclear species by assuming a plausible continuous sequence for the conditions under which they formed.

Similar difficulties arise if the assumption is made that the chemical elements formed from the disintegration of an immensely large "polyneutron". This so-called polyneutron fission theory was proposed by Mayer and Teller in 1949. Unfortunately, the theory only vaguely approximates isotopic compositions of the medium-heavy elements, and the overall ratio of the amounts of heavy to light elements is not in agreement with that observed in nature.

All these early attempts to understand the way in which the elements formed led to the conclusion that the heavy elements for which $A > 70$ are more promising as a means of leading to new information than the light ones that constitute most of the matter of the universe. This is because of the relatively large number of isotopes and the relatively well-known abundance ratios of the trace elements in meteorites, as compared to the main constituents of the universe.

2.2 THE SOLAR SYSTEM ELEMENTS AS A MIXTURE OF DIFFERENT COMPONENTS AND THE INTERPRETATION OF THE ABUNDANCE RULES

Ten years after the rules governing the general abundance picture were established (Suess, 1947), a fundamentally new idea led to a new phase in their interpretation. This idea was first suggested by Burbridge, Burbridge, Fowler, and Hoyle (B²FH, 1956). It is based on the assumption that the matter surrounding us constitutes a mixture of several components, each of which has a different genetic history. Previously, the extreme constancy of the isotopic composition of the elements in our solar system had made this assumption appear implausible. However, there are now overwhelming arguments that it must be the right answer to this question of the origin of the elements. We must assume now that there are two main components comprising the matter in our solar system: one of them consisting of species formed by a slow neutron buildup and the other consisting of nuclear species formed by rapid neutron buildup. Two processes, the S-process and the R-process, are assumed to have occurred independently, under different conditions, and at different locations:

1. An R-process, or a rapid neutron buildup, takes place if the capture of neutrons by the individual nuclei is faster than the β decay of the newly formed daughter nucleus. This nucleus will then have a chance to capture an additional neutron, and the process will continue until the rate of neutron capture becomes slower with increasing neutron excess than the rate of β decay that increases with increasing neutron excess.

2. An S-process occurs when the concentration of ambient neutrons is so low that the time between capture of one neutron and the next is long enough to allow the resulting unstable nuclei to undergo β decay.

B^2FH suggest that these processes occur in the interior of stars, as can be observed today. S-processes take place in regular massive stars that tend to develop into novae, and R-processes are derived from even more massive stars that develop into supernovae. There can be no doubt now that these two types of processes must have occurred. The idea of B^2FH of taking the solar system's matter as representing a mixture of different components has opened up a wide field of astrophysical and cosmochemical research. The work of B^2FH has been most gratifying for cosmochemists, as for the first time results of trace element analyses of rocks and meteorites can be linked to primeval astrophysical processes. The trace element content of one type of meteorite, the type I carbonaceous chondrite, corresponds in a surprisingly perfect manner to the abundance rules proposed before (see Sect. 1.2.4). These rules, and hence the abundances of the individual nuclear species, can be explained in a straightforward manner by assuming that the solar system matter consists essentially of two components. It is now possible to estimate how much of each nuclear species was formed by the R-process and how much was formed by the S-process. The most conspicuous feature in this abundance distribution is the systematic difference in the abundances of even- and odd-mass-numbered species, the even–odd effect (see Sect. 1.2.2). Another conspicuous difference between even-A and odd-A species is the functional dependence of their abundances on mass numbers, which is more regular for odd-A nuclei than for even-A ones. For even-A species another feature was recognized in the heavier element region ($A > 70$). The isobar with the higher neutron excess number is usually more abundant (Suess, 1949) than the one with the lower neutron excess number. In other words, the species shielded against β decay are less abundant than the shielding species. Most nuclides that contain a magic number of 50, 82, and 126 neutrons are more abundant than their neighbors. These features correspond perfectly with the two components present in solar system elements. In the case of even-mass-numbered species, a shielding isobar may consist of R-material only or it may be a mixture of R- and S-material. The shielded isobars are S-material unless they are "excluded", that is, they cannot form by either one of these processes. Odd-A species, of course, may contain matter formed by both R- and S-processes.

These empirical observations can easily be understood by considering the difference in the character of the overall abundance distributions that can be expected to result from the two types of nuclear processes. The high neutron flux required by the R-process can only occur under extreme conditions, such as very high temperatures at which the individual properties of the nuclear species do not show marked differences. In particular, at high temperatures even- and odd-A species, present largely in their excited states, will not show systematic differences of their neutron capture cross sections. One may therefore expect little or no even–odd effects in the yields of R-processes.

The situation is different in the case of S-processes. Here, the rate of formation of stable species with mass number A will be

$$d[A] = f[A - 1]\sigma_{A-1}\, dt,$$

and their rate of consumption by neutron capture,

$$-d[A] = f[A]\sigma_A\, dt.$$

Here, $[A]$ denotes the abundance of an isobar with mass number A. Normally, only one isobar at each mass number will participate in an S-process. In rare cases in which two isobars contribute, the expression will be valid for each isobar separately. f is a function of flux and energies of the ambient neutrons and of temperature. σ_A is the neutron capture cross section of the respective nuclear species at the prevailing neutron energies. Assuming that f is constant for the whole mass range during the S-process and assuming a steady state in which the rate of production of a nuclear species is equal to its consumption, one obtains the following for the abundance $[A]$ of a nuclear species with atomic number A:

$$[A] \approx [A - 1]\sigma_{A-1}/\sigma_A.$$

To obtain a more quantitative expression for the S-process yields is complicated, but it is relatively easy to see that the product abundance $[A]$ multiplied by the (Maxwell averaged) neutron capture cross section σ_A should be a smooth function of mass number. A sufficient number of neutron capture cross sections of individual nuclear species have been determined experimentally to enable us to see to what extent this is true (Käppeler et al., 1982). This is shown in Figure 16 (Rule 1). The con-

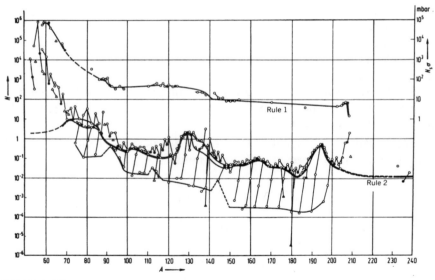

FIGURE 16 Logarithm of the abundances of the individual nuclear species after Palme et al. plotted against mass number. The heavy line marked "Rule 2" indicates abundances resulting from the R-process. In the upper part of the figure the line marked "Rule 1" indicates the products of residual S-process yields multiplied by neutron capture cross sections. (Courtesy of Professor H. D. Zeh.)

cepts of the R- and S-processes lead to a direct and convincing interpretation of the two most important abundance rules: first, that of a smooth dependence of nuclear abundances on the mass number, and second, that of a smooth dependence of the product of abundance multiplied by neutron capture cross section. The first rule can be recognized to hold in mass ranges where nuclides consist primarily of the R-component, as well as for the odd-A and "shielding" nuclear species that essentially are formed by the R-process. The second rule is valid for species that consist mainly of the S-component, such as "shielded" nuclear species.

The empirical observation that the abundance distribution of the odd-A nuclei is much more regular than that of the even-A species is explained by the fact that the odd-A species have a larger neutron capture cross section and hence a smaller S-process yield than the even-A species. The odd-A abundances have a larger R-process yield, and hence their abundance distribution shows a "smoother" mass dependence than the even species.

Perhaps the most direct evidence for neutron buildup processes comes from the effect of neutron shell closures upon the abundance distribution. For both the S- and R-process the neutron capture cross sections are important factors determining abundance yields. As expected, nuclides with closed neutron shells have exceptionally small neutron capture cross sections. Stable nuclear species on the S-path of neutron buildup and containing a magic number of neutrons, namely, $N = 50$, 82, and 126, all have an abnormally high abundance. For the case of the R-process, the cross sections of species in the unstable neutron-rich regions determine the ultimate abundances of the stable species with the corresponding mass numbers. Beta-unstable, high-neutron-excess species containing such magic neutron numbers, had formed in broad mass ranges that were smaller by about six to eight mass units than those of the stable species with magic neutron numbers. Correspondingly, broad abundance maxima exist, where A equals approximately 130 and 195. These two separate signatures of neutron buildup demonstrate that two different processes had indeed occurred. Products that might have formed under intermediate conditions are absent. The so-called "excluded" species on the β^+ side of the energy valley cannot form by either of these processes. The excluded species are present, however, only in relatively small amounts, about two orders of magnitude smaller than the other species within the same range of mass numbers. There are several ways in which these excluded species might have formed, for example, by spallation processes or by reactions with energetic protons.

This concept of considering our solar system matter as consisting of different components that formed independently allows an immediate interpretation of the most prominant abundance rules for the mass range $A > 70$: the lower abundances of odd-A nuclides compared to even nuclides are a consequence of the larger neutron capture cross sections of the odd-A species as compared to those of even-A species. As the product of S-component abundance times neutron capture cross section is nearly constant within a limited mass range, the S-component present in odd-A nuclides is correspondingly small. Therefore, odd-A nuclides of mass numbers $A > 70$ represent mainly R-component material. The even–odd effect of products formed by an R-process is expected to be much smaller than that of an S-process. Indeed, it can be seen that the R-component does not show any even–odd effect in its abundance distribution except in some limited mass ranges that follow neutron shell closures. The smooth dependence of abundances of odd-A species on their mass numbers (Rule 1) is explained in this way. The R-component predominates in odd-A species of most mass ranges.

For the case of the even-A abundances, a contribution of S-material can be recognized in most mass ranges. Exceptions are the two abundance humps around mass numbers 130 and 195. They consist essentially of R-component material (Fig. 15). The abundances of the even- and odd-A nuclear species are found to lie on the same abundance line. No even–odd effect can be recognized.

The two rules, (1) the smooth dependence of the R-component on mass number and (2) the smooth dependence of the product abundance times neutron capture cross section, can be used to estimate the composition of each nuclear species in terms of the two components in the medium-heavy and heavy mass ranges. The overall ratio of R- to S-component of the $A > 70$ mass range can be estimated to be about 3:1.

For the S-process, conditions such as neutron flux and energy can be calculated quantitatively if a preexisting abundance distribution of seed nuclei is assumed. In this way, S-processes in the interior of novae can be described by assuming plausible physical conditions. The parameters necessary for R-processes are more difficult to postulate. The only way to explain them by astronomically observable events appears to be to assume their occurrence during supernova explosions. It is, at present, the generally accepted hypothesis that debris from both nova and super-nova explosions accumulated into ordinary stars such as our Sun. An alternate scenario of R-process synthesis has been proposed by Amiet and Zeh (1971): the formation of heavy nuclei by slow neutron buildup under extremely high pressures. The nuclear reactions are easier to understand under such conditions; however, no cosmological model has as yet been proposed, that provides the required conditions.

It is generally assumed that R- and S-processes occurred completely independently. However, there are indications for a genetic relationship. In particular, it seems that the ratio of the amounts of R- and S-components does not change substantially over the whole mass range. This may be accidental, or it may be indicative of the presence of "seed nuclei" that were produced in some way by the R-process and then acted upon by an S-process.

The interpretation of the abundances of the light nuclear species ($A < 70$) is an astrophysical problem. These abundances are believed to be established during the lives of the individual stars. Investigations of whether solar wind-implanted gases in meteoritic and lunar materials have changed in chemical or isotopic composition during geologic times are now in progress. Otherwise, chemistry can contribute little to the interpretation of light element data.

One interesting problem that may be mentioned here is the high abundance of ^{56}Fe. The total mass of these nuclides present in our solar system is larger than that of all the other nuclear species from calcium to uranium combined. ^{56}Fe does not contain a magic number of neutrons, and it is generally assumed that it is formed in processes in the direction of thermodynamic equilibrium. However, in this mass range the binding energies of the nuclei show a flat maximum. At temperatures in question, a flat broad abundance maximum should be present and not a singularly sharp peak as manifested by the abundance of ^{56}Fe. Therefore, other kinetic processes may have contributed to the high abundance of ^{56}Fe. In particular, this is suggested by the fact that an isobar, ^{56}Ni (half-life 6.1 days) is "double magic"; it contains 28 neutrons and 28 protons. In any case, enough is now known about the origin of the elements that we might confidently say that work during the next few decades will lead to a clear understanding of the origin of the matter of which we consist and the matter surrounding us.

PART TWO

COSMOCHEMICAL PROCESSES

THREE

The Chemistry of the Primordial Mixture

3.1 ATOMS AND MOLECULES

3.1.1 Isotopic Anomalies

Perhaps the most important empirical observation so far discussed is the near-constancy of the isotopic composition of almost all the elements in practically all material investigated, (i.e., terrestrial as well as meteoritic and lunar.) This approximate constancy is also seen in the spectrum of sunlight and of reflected sunlight from the planets and other members of the solar system. Even comet tails show that their light is emitted by elements that are composed of basically the same isotopes.

On the basis of these measured isotopic compositions and of the relative abundances of the elements in the solar system, the firm conclusion was drawn that the individual elements constitute a mixture of materials from different sources with different genetic nuclear histories. The mixture consists of two main components, the R (or β^-) and the S (see Sect. 2.2), and one or two minor ones.

For several reasons the isotopic composition of all the elements in terrestrial and extraterrestrial materials must be expected to show at least some minor variations. The types of isotope variations to be expected are the following:

1. *Primeval isotope effects.* The materials with different nuclear genetic histories, in particular the two main components, have different isotopic compositions. One should expect that, the mixing of these components should not be completely perfect. With increasing accuracy of mass spectrometric measurements, minor variations in isotopic ratios consistent with the expected differences in the isotopic compositions of the genetic components should become apparent.

2. *Radiogenic isotope effects.* The genetic components must originally have contained radioactive nuclear species. Some of these may have been sufficiently long-lived to survive the period of time from their origin to the beginning of the chemical processes that led to the formation of the investigated objects. Any material that contains, or once contained, a radioactive species should also contain its daughter formed by radioactive decay, as long as the material represented a closed system and has remained chemically unchanged. If daughter isotopes are present, then the isotopic composition of the daughter element will vary, depending on the amount of the radiogenic daughter present. Lead and strontium, for example, are such elements. Their isotopic compositions vary in nature. The radiogenic daughters give information about the length of time an object can be considered to have been a closed system.

3. *Cosmogenic effects.* Not only the daughters of natural radioactive nuclear species, but also the so-called "cosmogenic" nuclear species occasionally affect the isotopic composition of certain elements. Energetic cosmic rays, colliding with other nuclides, lead to a breakup, or "spallation", of these nuclides. Fragments, which may or may not be radioactive, are produced as are occasional neutrons, which are soon absorbed by other nuclear species. Such "cosmogenic" nuclear species are present, for example, in the terrestrial atmosphere. Radioactive cosmogenic nuclides may give information about so-called exposure ages, or the time that has elapsed since the breakup of a large parent body.

The rate of formation of cosmogenic nuclides on Earth and in its vicinity is on the order of one atom per square centimeter per exposed surface per second—a negligible amount compared to the bulk of the elements normally present, except for the rare gases. These gases are present on Earth in amounts of only 10^{-11} to 10^{-8} parts of those adjacent elements in the Periodic Table.

4. *Chemical effects.* Isotopic species show differences in their chemical and physical properties. This is because the thermodynamic partition functions contain mass, as well as other properties, of the particles under consideration. In general, the differences in the mass-dependent constants of isotopes with mass m_1 and m_2 are functions of their mass difference and roughly proportional to

$$(m_1 - m_2)/(m_1 + m_2).$$

Of all elements, the hydrogen isotopes with mass values of 1, 2, and 3 (H, D, T) show the largest differences in their chemical properties. The partition functions also contain some mass-independent quantities (see, for example, Fowler and Guggenheim, 1939).

The differences in the chemical behavior of molecules containing different isotopes of an element, but otherwise of identical composition, can best be recognized by considering the constants for the thermodynamic equilibrium of an isotope exchange reaction of the type

$$XL + YG = XG + YL.$$

Here, X and Y denote arbitrary radicals and G and L heavy and light isotopes of an element, respectively. If the isotopes G and L of an element had exactly the same properties, then the equilibrium constant [XG][YL]/[XL][YG] would be exactly one. For most elements, only a slight deviation from unity is observed.

The difference in the ratio R of two isotopes of an element, given as the per-mil derivation δ from a standard abundance ratio R_{st}, is

$$\delta = [(R/R_{st}) - 1] \times 1000.$$

The most interesting observations would be those due to incomplete mixing of the two primordial components. In many cases, however, variations in isotopic composition may well be due to differences in chemical behavior.

3.1.1.1 Hydrogen

As expected, by far the greatest differences in chemical behavior are shown by the isotopes of hydrogen. When, in 1932, Urey discovered the heavy hydrogen isotope deuterium spectroscopically, he used the hydrogen obtained from an old apparatus that had been used to demonstrate the electrolytic decomposition of acidified water to college freshmen. Because the water had not been completely replaced for many years, the amount of deuterium in the gas was higher than in natural water, and this helped in the discovery of the isotope. Obviously, however, it also obscured the correct value for the deuterium concentration in water. Only when it was realized that the deuterium concentration in the electrolytically obtained hydrogen was depleted by nearly a factor of seven relative to the electrolyzed water was it possible to obtain an accurate value for the deuterium concentration in the surface waters on Earth. This value varies considerably because of the difference in vapor pressure of H_2O and HDO.

The mean D/H ratio in terrestrial ocean water is now taken to be 1.6×10^{-5}. Astronomical observations have shown that on Earth deuterium is enriched approximately eightfold compared with interstellar hydrogen. During the formation of the Sun, the deuterium appears to have been used up by nuclear reactions, and there is much less deuterium in the Sun than in interstellar and galactic space.

The outer planets—Jupiter, Saturn, Uranus, and Neptune—have large amounts of hydrogen in their atmospheres. This hydrogen is enriched in deuterium relative to interstellar galactic hydrogen, but less so than the hydrogen on the surface of the Earth. The deuterium content of the hydrogen in meteorites shows large variations (Boato, 1954). It is difficult to obtain exact values because of the presence of terrestrial water in most meteorites, but careful degassing of samples by stepwise heating often yields fractions that have a higher proportion of deuterium than the surface water of the Earth. Approximate values are given in Table VI.

The high deuterium content of hydrogen gas in the terrestrial atmosphere (containing $\sim 0.5 \times 10^{-6}$ parts of H_2) is remarkable, as it indicates that the H_2 is formed predominantly by photochemical reactions in the stratosphere (Barth and Suess, 1960) and not by biological processes on the Earth's surface. The high deuterium value, which is greater than that of seawater, is explained by the higher escape rate of H, as compared with D, from the upper atmosphere (Harteck and Suess, 1949). Table VI includes "pre-bomb" tritium values measured by Faltings and Harteck (1949) (see also Begemann, 1963), values that should be included

Table VI
Deuterium in the Solar System and Tritium on Earth

Material	$[D]/[H] \times 10^{-5}$	"Pre-Bomb" $[T]/[H] \times 10^{-18}$
Interstellar matter	~2	—
Interstellar molecules	100–1000	—
Solar wind	<1	—
Jovian planets	5–7	—
Earth		
Surface ocean water (SMOW)	16	~3 ÷ 10
Polar ice	8	0
Atmospheric H_2	20	~10^3
Meteorites	20–80	—

in discussions of the high deuterium content of molecular atmospheric hydrogen (Suess, 1953).

The fact that the Sun's hydrogen is greatly depleted of deuterium indicates that the members of the solar system formed before the main stage of deuterium-burning of the Sun, the so-called T-Tauri stage. The inner planets, their satellites, and the meteorites have formed after the loss of most of their hydrogen. Loss of hydrogen from a gravitational field leads to oxidizing conditions and also to deuterium enrichment in the residual hydrogen. This enrichment is in part due to the greater rate of loss of the light hydrogen isotope, as compared to deuterium by diffusion from the gravitational field (Sect. 3.2.2) and in part to exchange of hydrogen atoms with water vapor. Deuterium becomes enriched in the water that remains behind.

Calculations of the hydrogen isotope equilibrium constants are simplified by the fact that vibrational states are not excited at sufficiently low temperatures. At ordinary temperatures, the zero point energy $h\nu/2$ is the only state with vibrational energy to be considered in the calculations (Bigeleisen and Mayer, 1947).

The frequency, ν, and hence the zero point energy, is higher for the light isotope than for the heavy one. Hence, for the heavy isotope the compound will be energetically more favorable when the frequencies are lower. In equilibrium the heavy isotope is always enriched in the com-

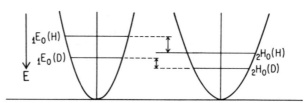

FIGURE 17 Binding forces as functions of atomic distance. A stronger force, and hence, higher vibrational frequencies prevail at left than at right. The difference in the zero-point energies is larger when the binding force is stronger, and this favors an enrichment of the heavy isotope. Under equilibrium conditions, the heavy isotope will be enriched in compounds where it is bound by stronger valence forces.

pound in which it is held by the stronger valence force, where vibrational frequencies are higher (Fig. 17). This should be true not only for hydrogen, but also for most other isotope exchange equilibria. It can also easily be seen that the differences in the reaction energies of isotope exchange reactions, and hence the differences in the chemical properties of the light and heavy isotopes, decrease with increasing temperature.

Figure 18 illustrates the temperature dependence of the deuterium exchange reactions $HDO + H_2 \rightleftharpoons H_2O + HD$ and $HDS + H_2O \rightleftharpoons H_2S + HD$ in the gas phase. Remarkably, the first data (crosses) for the deuterium exchange equilibria between H_2 and H_2O showed discrepancies in the experimentally obtained values compared to those calculated from statistical thermodynamics (Farkas and Farkas, 1934). Later measurements showed that the calculated results (the straight lines in Fig. 18) were more accurate than the experimental measurements (indicated by crosses) of these researchers.

It should be mentioned that the exchange of hydrogen isotopes between hydrogen gas and other compounds requires catalysts such as platinum. Exchange between H_2S and H_2O, of course, proceeds fast via hydrogen ions. This makes it possible to use this equilibrium for industrial enrichment of deuterium for the production of D_2O, the so-called "heavy water".

3.1.1.2 Oxygen

Particularly interesting are the isotopes of oxygen. As expected, $^{18}O/^{16}O$ ratios vary considerably in meteorites, just as on the surface of the Earth.

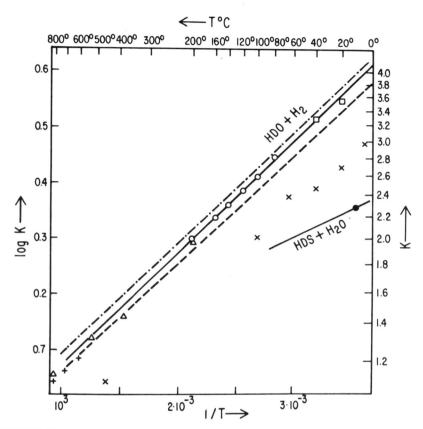

FIGURE 18 Decadic logarithms of the equilibrium constants for the gas reaction HDO + $H_2 \rightleftharpoons H_2O$ + HD is plotted as a function of $1/T$. Broken lines are calculated, points are experimental values. The line for the technically important equilibrium HDS + H_2O $\rightleftharpoons H_2S$ + HDO with one measured point is also calculated (Suess, 1949b).

Oxygen has three stable isotopes: ^{16}O, ^{17}O, and ^{18}O. They are present in ocean water in ratios of $99.76 : 0.037 : 0.204$. Evaporation and condensation processes lead to a $^{17}O/^{16}O$ fractionation half as large (or more accurately 0.52 times as large) as that of $^{18}O/^{16}O$ (Clayton and Mayeda, 1977). As expected, it was found that on the surface of the Earth the isotope ratios of oxygen in practically all the material containing oxygen—oxides, silicates, carbonates, etc.—showed variations for which $\delta^{18}O \simeq 2\delta^{17}O$ (neglecting second-order effects for cases in which $[^{16}O]$

is not much higher than [^{17}O] and [^{18}O], as always on the Earth's surface). Surprisingly, however, this is not always the case for the oxygen in meteorites. Clayton et al. (1976), discovered that some meteorites contain oxygen depleted to the same degree of ^{17}O and ^{18}O, as if several percent of an oxygen component essentially free of isotopes 17 and 18 were present. This has been considered irrefutable proof of the presence of a different genetic oxygen component.

However, such an interpretation met with severe difficulties, as no other element could be discovered in extraterrestrial material that had comparably large deviations in its isotopic composition as oxygen. A surprising discovery was made, however, by Thiemens and Heidenreich (1983), who showed that in the laboratory an electric high frequency discharge through O_2 at low pressures leads to the formation of isotopically abnormal ozone. This ozone shows a "mass-independent isotope fractionation" (Fig. 19) by which both heavy oxygen-isotopes are fractionated to the same degree relative to ^{16}O, the main isotope (δ^{17}O $\simeq$ δ^{18}O). An abnormal isotopic oxygen composition can thus be obtained in the laboratory by using a Tesla coil. Therefore, it is not necessary to invoke a supernova to explain the presence of abnormal oxygen in nature, even though in both cases the chemical mechanism that does lead to its formation is unknown.

3.1.1.3 Further Examples of Isotopic Anomalies

A very detailed review of the work so far carried out on isotopic anomalies of light elements was published by Pillinger (1984). The review clearly describes the many open questions that these anomalies in meteorites present. Selected examples were discussed in the preceding sections. Other elements could be added, although their isotopic ratios may be more important for the study of geochemical processes on the surface of the Earth than for the study of cosmochemical events. The ^{13}C : ^{12}C ratio in terrestrial material varies by up to about 30‰, depending upon whether the substance is organic or inorganic and whether the carbon is a carbonate or an organic compound. Interestingly, the ^{13}C : ^{12}C ratio in plant material depends on the photosynthetic pathway. C_3 plants give δ^{13}C $\simeq$ -25‰, and C_4 plants about -10‰ relative to the marine carbonate standard. Much larger variations have been observed in so-called FUN meteoritic inclusions (Wasserburg et al., 1977). The ^{15}N : ^{14}N

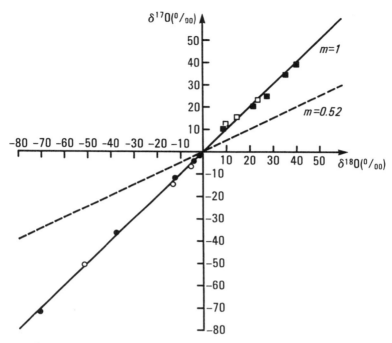

FIGURE 19 Variations of the isotopic composition of oxygen on Earth (broken line) and in photochemically produced ozone (squares) and in the residual oxygen (circles) are shown (Thiemens and Heidenreich, 1983). The solid line corresponds to non-mass-dependent isotopic fractionation ($\delta^{17}O = \delta^{18}O$), the broken line to mass-dependent fractionation [$\delta^{17}O = 0.5\ \delta^{18}O$ (or more accurately, $0.52\ \delta^{18}O$)]. (Courtesy of Professor M. H. Thiemens.)

ratios vary in a similar manner. There are indications that solar wind nitrogen may have changed its isotopic composition slightly over geologic time, an interesting observation that requires further study.

There exist many other experimental observations of isotope anomalies that elude any simple explanation. For example, neon in Allende meteorite inclusions appears to contain pure ^{22}Ne (Eberhardt, 1974). The only simple way to explain this would be to assume that this gas is the daughter of ^{22}Na (half-life, 2.6 years). But how could mineral formation occur only a few years after ^{22}Na formed by nuclear reactions? The case of excess ^{26}Mg may be easier to explain, as it is the daughter of ^{26}Al (half-life, 7.4 × 10^5 years). However, Esat and colleagues (1985) recently found that distillation of magnesium can lead to substantial mass fractionation. They believe that it is not necessary to invoke ^{26}Al decay to explain the ^{26}Mg anomaly.

In any case, there is no reason at this time to abandon the basic contention of cosmochemistry that our solar system originated from a well-mixed gas mass of essentially chemically and isotopically uniform composition. However, no two different cosmic gas masses can become ideally homogeneous during a finite time interval. Whether isotopic anomalies can be recognized depends on the sensitivity of the experimental instrument. The fact remains that, with the exception of light elements that participate in nuclear reactions on the Sun, the elements of the solar system by and large do not exhibit isotopic variations large enough to prove conclusively that they could not have resulted from ordinary chemical and physical processes. In rare cases, however, we do find indications that point to primeval effects from incomplete mixing of genetic components.

A large number of elements show small variations, on the order of 1 part in 10,000, in their isotopic compositions in peculiar inclusions in some types of meteorites, primarily in carbonaceous chondrites. A small number of elements do show relatively large deviations from normal isotope ratios, in particular, some "refractory" elements (see Sect. 3.2.1) such as titanium. For example, an enrichment of over 10% of ^{50}Ti (Fahey et al., 1985) relative to the other titanium isotopes has been found in a mineral (hibonite) in refractory inclusions of the Murrey C-2 chondrite. It has been suggested (Lugmair, personal communication, 1985) that the relatively large isotopic anomalies (Shima, 1985) in refractory inclusions were caused by their very early condensation from a gas in which the bulk of the two genetic components R and S was still gaseous but had not yet completely mixed. In this way, a small part of some refractory elements, being solids, had not participated in a last phase of the mixing process of the R- and S-components (see Sect. 1.2).

3.1.1.4 Rare Gases

The rare, or noble, gases, because of their chemical inertness, can almost be considered as isotopes of one and the same element, but with mass numbers ranging from 3 to 132 and beyond, if radioactive species are included. Rare gases do not form compounds that occur in nature; therefore, it might be thought that they are uninteresting from a cosmochemical viewpoint. However, this is certainly not the case. Even though their chemical behavior is simpler than that of any other element, their distribution is no less difficult to interpret. In nature they are frequently found fractionated by large factors relative to their original solar proportions. Their isotopic compositions vary by factors larger than those of any other element. Much work has been done to determine the relative amounts of rare gases and their isotopic compositions in the Sun, planetary atmospheres, meteorites, and other extraterrestrial objects.

On Earth, ^{4}He and ^{40}Ar are radioactive decay products. ^{4}He forms by α decay and ^{40}Ar from ^{40}K by K-capture. About 91% of ^{40}K decay leads to ^{40}Ca and 9% to ^{40}Ar. ^{3}He is largely a spallation product, formed in part from tritium decay. The solar system abundances of the rare gases cannot be determined in the same way as those of the majority of other elements. Fraunhofer absorption lines of rare gases are absent from the solar spectrum. For a long time, the only data available for the abundances of rare gases were derived from emission spectra of so-called planetary nebulae and from blue O and B stars with high surface temperatures. The spectrum of τ-Scorpii, carefully analyzed by Unsöld (1944, 1948), yielded a relatively high abundance for neon, close to that for oxygen. Later analyses gave neon values more than factor of two lower (Aller, 1961). Isotope systematics (Sect. 1.2.3) allow reliable estimates for the solar system abundances of krypton and xenon; estimates for the abundances of argon and neon are less accurate. The helium value is derived from the hydrogen–helium ratio, but unfortunately the hydrogen value, namely, the abundance ratio of [H] to [Si], is not as accurate as for most other elements. The best values for these elements, and also for the heavier rare gases, were obtained by Swiss workers (Geiss and Reeves, 1981) from solar wind data.

The relative amounts of rare gases in the terrestrial atmosphere are quite different. The ratios of neon to krypton and xenon in air are more than a 1000 times smaller than those of these elements in the Sun. Abundance ratios of the rare gases in the terrestrial atmosphere and the solar system were discussed independently by Harrison Brown (1948) and Suess (1949c) in detail. Obviously, only a small fraction of the gases originally associated with the condensed matter that now makes up the mass of the Earth has been retained in the terrestrial atmosphere. In

Figure 20 this fraction of each rare gas is plotted as a function of the atomic weight of its main isotope. Table VII lists the respective values.

The data show that an enormously efficient mechanism must have separated the gases from condensed matter before and during the formation of the Earth and the inner planets. The Earth's atmosphere has retained less than 10^{-10} parts of the neon and less than 10^{-6} parts of the krypton and xenon originally associated with the terrestrial mass. In other words, neon has been separated from the heavy rare gases by a factor of about 10,000. The mechanism that caused the separation is not known, though it seems probable that surface adsorption of the gases has been important. This is suggested by the rare gas content of meteorites, which contain so-called planetary rare gases in proportions similar to those of the terrestrial atmosphere, as discussed below.

In 1953 Professor Cherdinchev at the University of Moscow noticed

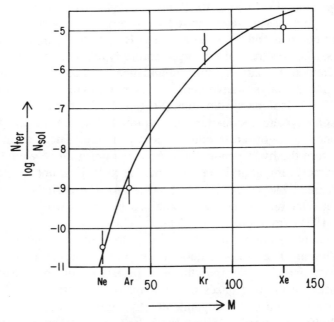

FIGURE 20 Amounts of rare gases in the terrestrial reservoirs relative to their solar abundances, normalized to the amount of silicon in the Earth (assumed to be 15% of its total mass), plotted versus the atomic weight of their main isotope.

Table VII
Decadic Logarithms of the Relative Abundances of Nonradiogenic
Rare Gases in the Terrestrial Atmosphere and in Gas-Rich Meteorites

Rare Gas	M	$\log N_{ter}$	$\log N_{sol}$	$\log$ Ratio
Ne	20.2	−4.0	+6.5	−10.5
Ar	36.3	−3.7	+5.0	−8.7
Kr	83.8	−5.2	+1.8	−7.0
Xe	131.3	−6.3	+1.0	−7.3

The values, denoted $\log N_{ter}$ and $\log N_{sol}$, respectively, are given relative to $\log N(Si) \equiv 6$ and are plotted in Figure 20. The amount of silicon in the Earth and in the meteorites is assumed to be 15% of the total mass. ($\log N_{sol}$ from Marti et al. 1972.)

that one meteorite in his collection, the Sekote Alin, contained a relatively large amount of gases that could not be analyzed by his ordinary technique, and he asked his colleague in Leningrad, Professor Gerling, to determine the composition of the gas by mass spectrometry. To their surprise, Gerling and his coworker Lefskii (1956) found that the sample consisted of rare gases, mainly helium. A search in Germany for meteorites containing rare gases then led to the discovery of several other so-called gas-rich meteorites. These meteorites did not belong to one particular class, but were found in several types of ordinary chondrites, such as Pantar and Breitscheid, and also in some achondrites, such as Kapoeta and Pesianoia. What these gas-rich meteorites did have in common was that they all appeared to consist of two kinds of material, a darker and a lighter one. Only the dark part contained high amounts of rare gases. Otherwise, no significant difference in the chemical composition of these two phases could be detected.

Various suggestions were made to explain how the rare gases had been incorporated into the meteoritic material, and some surprising discoveries were made. For example, Merryhue et al. (1962) showed that the heavy rare gases were homogeneously distributed in the individual chondrules, whereas helium and neon were enriched on the surface of the mineral grains and were also present in metal particles, in which solubility was negligibly small. On the basis of these facts and on the observations in Mainz, Wänke concluded that at least part of the rare gases had been implanted by solar wind particle radiation into the surfaces of meteoritic grains (see Suess and Wänke, 1965). As was shown by Signer and Suess,

(1963) the gases in gas-rich meteorites consisted essentially of two components, one with abundance ratios close to those of the rare gases in the solar wind and a second containing rare gases in ratios similar to those in the atmosphere of Earth, and presumably also of Mars and Venus. Accordingly, Signer and Suess suggested that the solar wind-implanted rare gases in meteorites be called the "solar rare gases" (or solar-type rare gases) and that the other rare gases, resembling, in their relative amounts, those occurring in the planetary atmospheres, be called the "planetary rare gases" (or fractionated rare gases) (Fig. 21).

This concept is supported by the fact that the two rare gas components in gas-rich meteorites show differences in their isotopic composition that are similar to those observed in solar wind and atmospheric rare gases. In meteorites the presence of additional components has to be considered, so that the following kinds of rare gases are present:

1. Solar rare gases, i.e., solar wind-implanted gases;
2. Planetary rare gases incorporated (presumably by adsorption) from ambient gases present during the accretion of meteorites;
3. Cosmogenic rare gases, or spallation products generated by interaction with highly energetic cosmic rays;
4. Radiogenic gases, mainly ^{4}He and ^{40}Ar, and ^{129}Xe that formed by the decay of extinct ^{129}I.

The huge differences in the relative amounts of rare gases on Earth compared to those in solar wind (namely, more than a factor of 10,000 in the ratio of neon to krypton and xenon) suggested large variations in isotopic composition of these gases. Measurements of their isotopic compositions were soon carried out at a number of research centers.

Neon and xenon show the largest differences in isotropic composition of the solar and planetary gases. The ^{20}Ne/^{22}Ne ratio is ~ 12 in the solar gas and ~ 9 in the planetary gas. This could be interpreted as a mass-dependent effect of the same order as the effect that separated rare gases from each other, but no corresponding effect can be recognized in the isotopic composition of the other noble gases. Highly anomalous components of neon and xenon were discovered in meteorites (Eberhardt, 1974; Marti, 1969), components that cannot be explained in any other way than as the result of nuclear reactions shortly before or during the formation of the solar system.

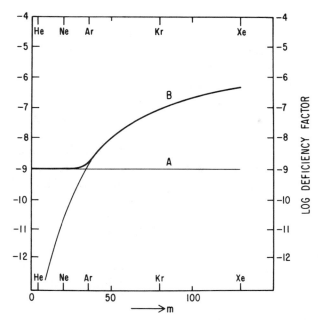

FIGURE 21 Rare gas content of gas-rich meteorites shown schematically as a function of their atomic weight to illustrate that they constitute two components, a component A, called "solar rare gases", and a component B, called "planetary rare gases". (Signer and Suess, 1963.)

A special case is ^{129}I (half-life, 17×10^6 years). There can be no doubt that excess ^{129}Xe found in meteorites is the product of the now extinct ^{129}I. Its occurrence, discovered by Reynolds (1963), gave rise to a new field of isotope research called xenology.

Radiogenic rare gases help in the determination of meteorite ages. The ages agree frequently, but not always, with the time elapsed since the formation of the solar system. Cosmogenic rare gases help determine the time elapsed since the break-up of parent bodies into the smaller objects that constitute the present meteorites. The planetary rare gases apparently have a complex history.

We know that helium, like hydrogen, escapes from the terrestrial atmosphere. The residence time of helium in the atmosphere is on the order of 10^6 to 10^8 years. This value is very sensitive to the temperature at the top of the atmosphere, and this temperature changes with the state of the Sun. Also, gas ions have higher escape velocities as they follow the escaping electrons, because of their coulomb forces.

The abundances of the two helium isotopes in the terrestrial atmosphere are entirely unrelated to their abundances in the Sun and solar system. The ^{3}He:^{4}He ratio in the atmosphere is only $\sim 10^{-7}$, whereas it is on the order of 3×10^{-4} in the present solar wind. The ^{3}He in the terrestrial atmosphere is in part derived from solar wind and in part cosmogenic, either directly or via tritium, which decays with a half-life of 12 years to ^{3}He. It has also been suggested that primordial ^{3}He diffuses from the interior of the Earth through the ocean floor together with primordial and radiogenic ^{4}He (see Suess and Wänke, 1965).

The ^{4}He content of the atmosphere mainly represents a steady state between helium released from radioactive uranium- and thorium-containing surface rocks of the Earth and the escape of the helium into interplanetary space. The escape of neon, which is five times as heavy as ^{4}He, is negligible, even if partial ionization into neon ions and electrons is considered. The influx of helium from the Sun may not be negligible over the course of geologic time; it has to be considered in quantitative estimates.

The ^{40}Ar in our atmosphere, which is at least 300 times more abundant than what would be expected without any radiogenic argon, must have come from several different sources: (1) from potassium-containing material during the accretion of the Earth; (2) from volcanic exhalations; (3) by weathering of potassium-containing rocks during the geological periods; and (4) from potassium decay in the oceans.

It is interesting that the krypton-to-xenon ratio in air is considerably greater than in meteorites. This indicates that some of the xenon associated with nonvolatile terrestrial material is present in the interior of the Earth, whereas krypton and the lighter rare gases largely found their way into the atmosphere.

Begemann (see Göbel et al., 1978) pointed out that the depletion coefficients of the planetary rare gases are a simple function of their ionization potentials. The higher the ionization potential (highest for helium—about 24.5 V), the smaller the fraction of the rare gas that has remained in the terrestrial atmosphere. However, there is also a correlation of ionization potential with the constants for the Freundlich adsorption isotherms of rare gases. The possibility that all the rare gases were retained with the solids through adsorption on elementary carbon and other substances with large surface areas has been discussed frequently. Begemann's correlation certainly reopens the discussion of these possibilities.

3.2 THE CONDENSATES

3.2.1 The Condensation Sequence

With the exception of a minute fraction (Sect. 3.1.1.3), the material that surrounds us shows no measurable indications of incomplete mixing of genetic components. This can only be explained by assuming that both the R- and the S-components were gaseous when the mixing occurred. At some time, a practically homogeneous gas mass must have existed with a temperature sufficiently high (higher than $\sim 2000\,°$K) and a total gas pressure sufficiently low that no condensed matter was present. With decreasing temperature and increasing pressure, substances with sufficiently low vapor pressures, first the so-called refractory substances, began to condense. Condensation of materials of relatively low volatility, the so-called moderately volatiles, and finally the volatile material followed. In this way we have a condensation sequence that can be calculated from thermodynamic data. Is this sequence reflected in any kind of meteorite?

The elemental condensation sequence, that is, a sequence that should be reflected in the condensates, will depend not only on temperature and pressure, but also on the main chemical composition of the gas, in particular, its oxygen content, or, more accurately, its thermodynamic oxygen fugacity. This oxygen fugacity is approximately proportional to the H_2O/H_2 ratio, and also to the CO_2/CO ratio. In order to calculate these ratios, we must know the elemental abundances of hydrogen and oxygen, and of the elements that combine with oxygen under the prevailing conditions. Complications arise in connection with the oxides of iron and carbon. Abundances of other elements, such as sulfur, are too low to affect the oxygen activity appreciably. At very low oxygen fugacities, sulfur replaces oxygen in many meteoritic minerals (Sect. 4.1.4). Table VIII shows the condensation sequence according to Grossman and Larimer (1974) for a gas of solar composition and a total pressure of 10^{-3} atm (before loss of H_2).

The condensation sequence changes with pressure and temperature.

Also, many metals, for example, the alkali earth metals, are much more volatile than their oxides or silicates. The opposite, however, is true for many other elements. For example, in the presence of a sufficiently high partial pressure of hydrogen, Al_2O_3 will form AlO or other suboxides, all more volatile than the most stable aluminum oxide:

$$Al_2O_3(s) + H_2(g) = 2AlO(g) + H_2O(g).$$

The same is true for many other refractory elements. At higher H_2 pressures, the redoxy equilibrium is shifted to form suboxides and metal with higher vapor pressure than the fully oxidized compounds. Silicon is another element that becomes more volatile at higher H_2 pressures. Its dioxide, quartz, has a relatively low volatility, but under reducing conditions it will form SiO, a gas; or at pressures prevailing on Jupiter, for example, it will react with excess hydrogen to form gaseous SiH_4 and H_2O (Eucken, 1944):

$$SiO_2 + 4H_2 = SiH_4 + 2H_2O.$$

In a gas of solar composition at relatively high pressures and temperatures, liquid metallic iron is the first major constituent that condenses with decreasing temperature (Anderson, 1973); nickel and "compatible" trace elements will condense with it. (Compatible elements are those that fit into the crystal structure of the main constituents.) Liquid metal will form at pressures at which the condensation temperature exceeds the approximate triple point temperature of pure iron ($\simeq 1800°K$). The total gas pressure, including H_2 in solar proportions, will then be about 1 atm. Magnesium silicates follow at a somewhat lower temperature of about 1700°K. These temperatures are considerably higher than those given in Table VIII, where all are below the triple point of the condensates. However, it can easily be seen that, with the exception of the carbonaceous chondrites, large parts of most meteorites, and in particular the chondrules must have been molten before or at the time they obtained their final structure. Direct condensation could therefore only have occurred from a gas of considerably higher temperatures and pressures than given

TABLE VIII
Condensation Sequence After Grossman and Larimer (1974) of
Elements in Gas of Solar Composition and a Total Pressure of 10^{-3}
atm

Temp. (°K)	Lithophiles	Siderophiles	Chalcophiles
> 1600	**Refractories** *Oxidized:* Al, Ca, Ti, Be, Sc, V, Sr, Y, Zr, Nb, Mo, Ba, La and REE, Hf, Ta, Th, U		
≃ 1450	*In magnesium silicates:* Mg, Fe, Si, Cr, Li, Mn	*As metals:* Fe, Ni, Co, Cu, Ge, Pd, Au, As, P, Os, Ir, Ru, Rh	
≃ 1000	**Moderately volatiles** F *Oxidized:* Na, K, Rb, Cs, Zn		
< 700	**Volatiles** *Halides:* Cl, Br, I		*As oxides or sulfides:* Ag, Sn, Sb, Se, Te, Tl, Cd, In, Hg, Pb Bi

Atmophile compounds and their boiling points in °K at 1 atm.: H_2O: 373.2; HCN: 299.2; SO_2: 263.2; NH_3: 240.2; COS: 223.5; CO_2: 195.2; C_2H_2: 189.6; C_2H_4: 184.6; Xe: 165.2; CH_4: 111.8; O_2: 90.3; Ar: 87.3; CO: 81.9; N_2: 77.4; Ne: 27.2; He: 4.2.

in Table VIII. This excludes the possibility of chondrule formation (in a fiery rain) by direct condensation at pressures of $\simeq 10^{-3}$ atm.

3.2.2 Hydrogen Loss

During the formation of the terrestrial planets, volatile material separated from planetary matter, except for very small fractions (10^{-11} parts) of the original gases that constitute the planetary atmospheres. Hydrogen and helium were lost more rapidly than heavier gases, including water vapor and CH_4, during the time the planetary masses were accumulating. Hydrogen continued to escape throughout geologic time to the present (Poole, 1941; Harteck and Jensen, 1948).

Free hydrogen in the atmosphere is produced through photolysis of water vapor in the upper atmosphere (Barth and Suess, 1960). The present rate of escape of hydrogen from the top of the terrestrial atmosphere is on the order of 3×10^8 atoms/cm^2 of Earth's surface/sec (Hunten, 1973).

The rate of escape of hydrogen from a gravitational field can be calculated from the speed distribution function of the gas molecules and from the velocity at the "exosphere," the gas layer from which the molecules escape without further collisions (Jeans, 1928). We are interested here primarily in the dependence of the escape rate on particle mass, temperature, and gravity at the exosphere. In cases where the escape from the exosphere is fast compared to the upward gas transport through the atmosphere, the exchange rate depends on this transport rate and is approximately proportional to $m^{-1/2}$, with m the mass of the gas particles. In cases where the escape from the exosphere is slow, the rate will be approximately proportional to $\exp(-mG/kT)$ with mG the potential and kT the kinetic energy of the escaping particle. An exact expression for the escape rate was obtained by Jeans (1928) with the assumption that the escaping particles maintain a Maxwellian speed distribution. Spitzer (1952) and others after him obtained more rigorous approximations. A detailed discussion of these results can be found in the book by Lewis and Prinn (1984).

All the members of the solar system with orbits closer to the sun than Jupiter have lost their elementary hydrogen. Objects the size of the Moon and smaller have also lost all gaseous substances, including all water vapor. On the basis of the amounts of neon in their atmospheres, this presumably occurred during T-Tauri stages of the Sun. Thereafter, new atmospheres and hydrospheres formed from H_2O and gases that reached the surface from the interior of the planets, where temperatures were rising.

The present atmosphere of the Earth contains a small fraction of molecular hydrogen, H_2, in its atmosphere, approximately 0.5×10^{-6} parts. As stated above most of this hydrogen is formed by photolysis of water vapor in the atmosphere. A small part of it presumably comes from biological decomposition of organic substances, and also from anthropogenic sources (see. 3.1.1.).

The loss of hydrogen from an Earth-like planet must have occurred in several stages. At first, hydrogen gas, retained together with other gases, left rather rapidly; then hydrides such as NH_3, PH_3, H_2S, CH_4, and H_2O were decomposed photochemically by ultraviolet sunlight. The hydrogen formed in this way left the gravitational field of Earth. It was long assumed that these processes became ineffective as soon as traces of oxygen became present. The reaction $H_2O + h\nu = H + OH$ yields atomic hydrogen, which quickly reacts with O_2 to give HO_2 and finally O_2 and H_2O. However, a small fraction of the photolysis of H_2O ($\approx 5\%$, depending on the ultraviolet spectrum) reacts to $H_2 + O$. Molecular hydrogen, H_2, is relatively stable and is the main source of hydrogen in the atmosphere (Barth and Suess, 1960).

It is often assumed that a large part of the 0.5×10^{-6} parts of H_2 in the atmosphere has a biochemical origin. That this is not the case can be recognized from the deuterium content of this hydrogen (Harteck and Suess, 1949). Biochemical hydrogen has a deuterium content close to the equilibrium concentration, which is four times lower than that of the ambient water (see Sect 3.1.1.1). The deuterium content of the molecular hydrogen in air, however, is greater than that of atmospheric water vapor and not at all in thermodynamic equilibrium at the prevailing temperatures. Biochemical processes yield hydrogen in isotopic equilibrium with water, whereas photochemical processes do not fractionate isotopes appreciably.

The higher deuterium content in the molecular hydrogen as compared to the ambient water vapor can be explained by the more rapid escape of the light isotope from the exosphere. When comparing the hydrogen escape rate with the rates for other gases, for example, helium, it may be worth remembering that ionized gas, consisting of ions and electrons, escapes at approximately the same rate as a gas with half its molecular mass, or, more accurately, with the average mass of the two particles tied together by coulomb forces.

Hydrogen loss from the atmospheres of the terrestrial plants has led to oxidation of the carbon present on the surface to CO_2 (see Sect. 3.2.3). It led to the formation of traces of free oxygen even before plant life produced the large amounts of free oxygen now present. Loss of hydrogen must also have been an important factor in the formation of highly oxidized substances such as nitrates in carbonaceous chondrites.

3.2.3 The Cosmochemistry of Carbon and the Formation of Prebiotic Substances

A special case is carbon. It is not a condensate in the usual sense, as, in thermodynamic equilibrium with a gas of solar composition and at ordinary temperatures, this element will always be present as a gas: under reducing conditions as CH_4 or another hydrocarbon, and under oxidizing conditions as CO_2 or CO. Nonvolatile polymers and elementary carbon in the form of graphite, soot, or diamonds are extremely inert; once formed they do not react easily to give a gaseous compound. However, they form readily in radical and charged particle reactions; an electric discharge decomposes CH_4 under certain conditions almost quantitatively into carbon black and hydrogen. Alfven and Arrhenius (1976) have discussed in detail the role of "plasma", namely ionized gases, in the early solar system. Such plasma appears to be a plausible medium for the formation of elementary carbon. But atomic and radical reactions can also lead to its formation. For example, the following processes lead to the formation of carbon black or to substances found in carbonaceous chondrites:

$$CH_4 + H \rightarrow CH_3 + H_2$$
$$CH_3 + H \rightarrow CH_2 + H_2$$
$$\vdots \qquad \qquad \vdots$$
$$CH + H \rightarrow C + H_2.$$

Likewise, there are many other ways by which carbon compounds can condense into refractory materials. At the lower temperatures ($< 400°K$) of the atmosphere of primitive Earth, equilibrium conditions may well have existed that led to the formation of a variety of simple organic compounds, especially in the presence of liquid H_2O. This is an important problem in connection with the origin of organic material. The following "thought experiment" elucidates this in a constructive manner.

Continuous escape of hydrogen from a system containing H_2, CH_4, and H_2O will lead to the formation of CO and CO_2 in increasing amounts.

Finally, with the loss of all the hydrogen, part or all of the carbon will be in the form of its monoxide or dioxide, and either some carbon or some oxygen will be left in its elementary state, depending on the initial carbon-to-oxygen ratio in the system.

If, as on the surface of the Earth, a large excess of H_2O is present, then an atmosphere containing free oxygen and CO_2 will develop. Before this state is reached and the oxidation of CH_4 completed, however, partially oxidized carbon compounds and higher hydrocarbons will be stable in thermodynamic equilibrium. At sufficiently high temperatures and ordinary pressures, CO is the only intermediate product in the process of oxidation of CH_4 to CO_2. At low enough temperatures, and especially in the presence of liquid H_2O and NH_2, photochemical oxidation of CH_4 leads to a great variety of organic compounds (Groth and Suess, 1938). The nature of these compounds depends largely upon the specific conditions. The gradual shift in the redoxy state of the system provides the free energy difference necessary for biological metabolism (Miller and Urey, 1959). Numerical calculation of the equilibrium concentrations of the various gases and of the exact conditions under which elementary carbon forms in thermodynamic equilbrium requires the solution of five partly nonlinear equations:

$$\Sigma H = 2[H_2] + 2[H_2O] + 4[CH_4] \tag{1}$$

$$\Sigma C = [CO_2] + [CO] + [CH_4] \tag{2}$$

$$\Sigma O = 2[CO_2] + [CO] + [H_2O] \tag{3}$$

$$K_1 = ([CO][H_2O])/([CO_2][H_2]) \tag{4}$$

$$K_2 = ([H_2]^4 [CO_2])/([CH_4][H_2O]^2). \tag{5}$$

ΣH, ΣO, and ΣC are the total amounts of hydrogen, oxygen, and carbon present in the system, expressed in atmospheres of monoatomic gas according to equations (1), (2), and (3).

In order to calculate the partial pressures of $[H_2]$, $[CH_4]$, $[CO]$, and $[CO_2]$ as functions of ΣH for various values of ΣO and $\Sigma C/\Sigma O$, Equations (1)–(5) were solved using the numerical values of the equilibrium constants K_1 and K_2 as given in Table IX (Suess, 1962).

In Figure 22 the concentrations of CH_4, CO, and CO_2, calculated from Equations (1)–(5), are shown as functions of ΣH at $T = 400°K$.

TABLE IX
Equilibrium Constants Calculated from Data in JANAF Table (1960)

T (°K)	K_1	K_2 (atm)^2	K_c (atm)
400	6.44×10^{-4}	3.63×10^{-13}	5×10^{-14}
600	3.52×10^{-2}	1.37×10^{-5}	1.7×10^{-6}
800	0.236	0.128	1×10^{-2}
1000	0.694	36.9	1.72
1400	2.15	2.96×10^4	5.7×10^2
2000	4.53	4.8×10^5	3.8×10^4

Figure 22 also shows the values of the ratio $[CO]^2/[CO_2]$. For the formation of solid carbon this ratio must be equal to the equilibrium constant K_c for the reaction $C + CO_2 = 2CO$ (Boudouard, 1901). The value K_c of this constant is indicated in the insert to Figure 22. No carbon forms at high temperatures in equilibrium during oxidation of CH_4 to CO_2. At 400° K, however, a narrow range of ΣH exists at which elementary carbon will be stable under these specific conditions, as is shown in the insert in Figure 22.

The purpose of the calculation was to find the conditions under which elementary carbon and organic compounds form in thermodynamic equilibrium. Figure 23 summarizes the pertinent information obtained from the calculations. If we start at the left side of the diagram, where all the carbon is in the form of CH_4, and decrease ΣH by removing hydrogen from the system, then the point in the diagram representing the chemical composition of the system will move horizontally to the right until it reaches the solid line for the respective temperature. At this point elementary carbon will become stable. In the gas phase, the carbon will be in the form of CH_4, CO, and CO_2. Removal of more hydrogen from the system will result in a decrease in the $\Sigma C/\Sigma O$ ratio in the gas phase because of the formation of solid carbon, and the point representing its composition will follow the solid curve until its minimum is reached. From then on, carbon will become oxidized until it is completely consumed and the original $\Sigma C/\Sigma O$ ratio in the gas phase is reached. The final state at which all the hydrogen has left will depend on the $\Sigma C/\Sigma O$ ratio. A sufficiently small $\Sigma C/\Sigma O$ ratio will lead to the formation of free O_2. The broken line in the lower part of Figure 23 gives the boundary

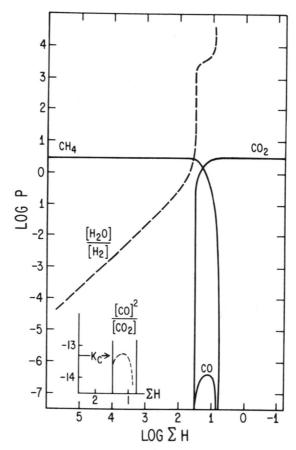

FIGURE 22 Decadic logarithms of the partial pressures in atmospheres of CH_4, CO, and CO_2 as a function of the logarithm of the total hydrogen in the system, as calculated from Equations (1)–(5) for $T = 400°K$ and a total oxygen pressure of $\Sigma O = 10$ atm. Also shown are the ratios $[CO]^2/[CO_2]$ and $[H_2O/H_2]$. (From Suess, 1962.)

for the presence of free oxygen in the gas phase. The broken line in the upper part of the diagram limits the area where solid carbon is present for stoichiometric reasons under all conditions. In general, given the presence of carbon and carbon compounds of low volatility in planetary objects, the C/O ratio in the solar system is an important quantity.

Our knowledge of this quantity is based entirely on astronomical observations. The most probable value of the C/O ratio in the Sun is ≈ 0.46

(Palme et al., 1981). This value and the diagram shown in Figure 23 suggest that it is plausible to assume that a fraction of carbon atoms present in the primordial terrestrial atmosphere found their way together with the condensing water into the oceans in the form of prebiotic organic matter. Optimal conditions for this would occur when, as a consequence of hydrogen loss, the system changed from a reducing to an oxidizing state, as in Figure 23 where the solid curve has its minimum. At relatively

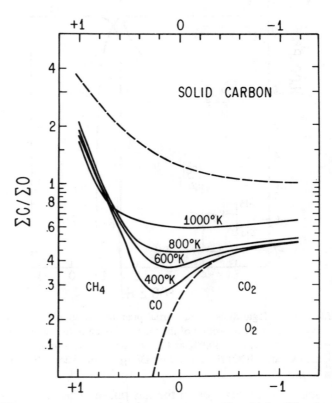

FIGURE 23 Phase diagram, showing the conditions under which solid carbon and many organic substances are stable in the presence of equilibrium amounts of CH_4, CO, CO_2, H_2O, and H_2. The solid lines show the boundaries above which solid carbon is stable for temperatures as indicated. Broken lines give the boundaries for the presence in equilibrium of O_2 and of solid carbon, respectively, under all conditions. The lines are calculated for a total oxygen pressure of $\Sigma O = 10$ atm, but the result is practically pressure-independent within several orders of magnitude. (From Suess, 1962.)

low temperatures, a variety of organic compounds such as aldehydes and alcohols will form instead of elementary carbon. This will always be the case when liquid H_2O is present. Nitrogen, sulfur, and phosphorus compounds will, of course, readily participate in the reactions, as described by Miller and Urey (1959).

FOUR

The Members of the Solar System

4.1 METEORITES

4.1.1 Classification of Meteorites

Since the time they formed from solar matter, meteorites have undergone less chemical change than any material on the surface of the Earth. In fact, radioactive age determinations show that no major changes have taken place in most meteorites since the solar system formed 4.6×10^9 years ago (see Part 5). The type of chemical fractionation that is conspicuously evident as having occurred during the formation of most of them is that of a separation into three main mineralogical phases: (1) a metal phase, (2) a sulfide phase, and (3) a silicate phase. Accordingly, Goldschmidt divided the chemical elements of the Periodic Table into three groups: siderophile, chalcophile, and lithophile elements (see Table VIII). He called a fourth group not present in meteorites atmophile elements. This classification, however, is often not rigorously defined, as many elements are distributed among two or three phases in varying proportions depending on the prevailing oxygen fugacity.

Close to 2000 individual meteorites are described in the literature (Prior and Hey, 1953). As early as 1843 it had been recognized (Partsch,

1843) that two fundamentally different types of objects fall from the sky: irons and stones. Since then, these objects have been classified into an increasing number of groups and subgroups. Only their main classes are listed here (Arnold and Suess, 1969). More recent subdivisions of these classes (Dodd, 1981; Wasson, 1985), mainly from a mineralogical viewpoint, are not discussed here, nor are the several thousand newly recovered meteorites from Antarctic ice.

I. Irons—Main phases:

kamacite (α-iron)
taenite (γ-iron)
troilite (FeS)
graphite (C)

Classification:

A. Nickel-rich ataxites (42 known): Generally nickel greater than 12%. Fine crystals.

B. Octahedrites (442 known): Nickel 6–12% (mostly 7–10%). Extended interleaved crystals of kamacite and taenite (Widmanstatten figures).

C. Hexahedrites (69 known): Nickel less than 6%. Mainly kamacite.

II. Stony-irons—Main phases: as above. Also:

olivine (Mg, Fe)$_2$SiO$_4$
pyroxene (Mg, Fe)SiO$_3$
feldspars (complex Na, Ca, K aluminosilicates)

Classification:

A. Pallasites (43 known): Mostly metal (exterior phase) like octahedrites. Olivine.

B. Mesosiderites (25 known): Metal, pyroxene, feldspar.

III. Stones—Main phases as above, with silicates predominant.

Classification:

A. Chondrites (1004 known): Contain chondrules, millimeter-scale spheroidal bodies of uncertain origin. All but carbonaceous have olivine, pyroxene, feldspar, metal phases, and troilite.

1. Enstatite chondrites (17 known): Highly reduced, all iron in metal and troilite.

2. Ordinary chondrites:

H-group (459 known). Less reduced, some iron in olivine and pyroxene also. Total iron about 28%.

L-group (494 known). Less iron and less metal than H-group. Total iron about 21%.
LL-group (67 known). Little free metal; total iron about 19%.

3. Carbonaceous chondrites (33 known): Contain organic matter, H_2O, hydrated minerals, little or no free metal: magnetite (Fe_3O_4), sulfate. Subdivided further into types I, II, and III, type I having highest organic and H_2O content. Total iron about the same as in the H-group.

B. Achondrites (69 known): Many classes, some with only one member. The more abundant classes are:
1. Aubrites–diogenites (calcium-poor; 16 known): Mainly pyroxene.
2. Eucrites–howardites (calcium-rich; 44 known): Feldspar and pyroxene.

A difficulty in computing the composition of the primordial solar matter from the elemental composition of stony meteorites arises from our ignorance of the relative amounts of the three phases in the parent bodies of meteorites. Meteorites that reach the surface of the Earth cannot easily serve as the basis for estimations of the relative amounts of mineralogical phases, as iron meteorites are better preserved during their fall and on the surface of the Earth than are stony meteorites and pallasites. The idea that the core of the Earth consists of metallic iron was derived from the high abundance of metal in meteorites. It led Tammann (1923) to his so-called "blast furnace model" of the Earth. Table X summarizes the estimates of early investigators.

TABLE X
Metal Content of the Planet Earth
Estimated by Different Investigators,
given as Parts by Weight Relative to the
Weight of the Silicates ≡ 100

Investigator	Metal	Sulfide
Noddack (1930)	68	9.8
Fersman (1934)	20	4
Goldschmidt (1938)	20	10
H. Brown (1949)	67	0
Urey (1952)	10.6	7

The values given by Harrison Brown (1949) and by Noddack (1930) are obtained from the ratio of the weight of the core of the Earth to its mantle, assuming that the average composition of the Earth is the same as the average composition of meteorites.

4.1.2. The Carbonaceous Chondrites

The carbonaceous chondrites, in particular the carbonaceous type I (C-1) chondrites, possess trace element contents that follow remarkably closely the nuclear abundance rules (see Sect. 1.2.4). This demonstrates conclusively that these meteorites contain the most primitive unaltered solid material that we can investigate in the laboratory. It can be shown that the bulk of material of which carbonaceous chondrites consist has never been heated above some 300°K since it accumulated into larger bodies. These chondrites, however, and in particular the type II and type III carbonaceous chondrites, contain high-temperature minerals, mainly olivine and pyroxene, frequently in the form of chondrules. The minerals must have been accreted from a different source than the bulk material, which contains H_2O and relatively volatile substances. This explains why the carbonaceous type 1 (C-1) chondrites contain trace elements practically in their primordial ratios, while the content of the main constituents—silicon, magnesium, and iron—varies. These elements exist in varying amounts in high-temperature minerals with low trace element content. The carbonaceous chondrites therefore consist largely of material that formed directly from the primordial nebula at relatively low temperatures and pressures but with the admixture of some high-temperature material that came from an area in the solar system closer to the Sun than their interstitial matter.

4.1.3 Ordinary Chondrites and Their Chondrules

About 20% of all the meteoritic matter reaching the Earth consists of chondrules, spherical objects 0.1 to 10 mm in diameter that contain the various chemical compounds present in stony meteorites. About 100 years ago a British mineralogist, Sorby, suggested that these spherical objects originated as a fiery rain that condensed out of solar gases and accumulated as hail stones. Together with some kind of dust, the so-called interstitial matter, they were compacted into what we now find in the form of stony meteorites. To consider these chrondrules as the first primordial condensates is indeed very attractive. The idea has been suggested many times during the last 100 years, although it is relatively easy to show that cooling solar matter cannot yield condensates with the chemical and mineralogical properties of chondrules (Suess, 1963).

The main constituents expected from cooling solar matter are, in addition to metallic iron, ferromagnesium silicates, particularly iron-containing olivine (Mg_2SiO_4) and pyroxene ($MgSiO_3$). The perfectly spherical form and the crystal structure of the minerals show conclusively that many, if not all of these chondrules, were molten droplets suspended in space at the time when they solidified at temperatures above the triple point of about 1600°K. In equilibrium, at this temperature, and in the presence of hydrogen in solar proportions, all of the iron should condense as metal. In order to obtain minerals containing oxidized iron, as present in ordinary chondrites, the hydrogen in the gas phase has to be depleted by more than a factor of 100 (Herdon and Suess, 1977).

In contrast to the carbonaceous chondrites, ordinary chondrites have elemental compositions that deviate markedly from that of solar matter. In particular, their iron and nickel (and also magnesium and silicon) contents differ to varying degrees from that of solar matter, allowing the ordinary chondrites to be classified into three groups, the Urey-Craig groups: H, L, and LL. Urey and Craig (1953) showed that the total iron (i.e., metallic, sulfidic, and oxidized iron), contrary to Prior's Law, is not constant in ordinary chondrites. H chondrites contain, by weight, about 28%, L chondrites about 22%, and LL chondrites (also called Soko-Banjites) about 20% total iron.

Prior's Law (1920) expresses what should be expected if the individual ordinary chondrites differed mainly by degree of oxidation. In that case, the total iron would be the same for all ordinary chondrites; that is, the oxidized iron would have formed "at the expense" of the iron in the metal, and the nickel content of the metal would have increased correspondingly. This is not the case, as can be seen from the many published diagrams in which metallic iron (plus sulfidic iron) content is plotted versus oxidized iron content (Fig. 24) (Keil and Fredriksson, 1964; Suess, 1964).

The iron content of the magnesium silicates is a most interesting property of meteorites. From a melt containing MgO, FeO, and SiO_2, the crystallizing olivine [orthosilicate, $(MgO)_{2-x}(FeO)_x \times SiO_2$] always

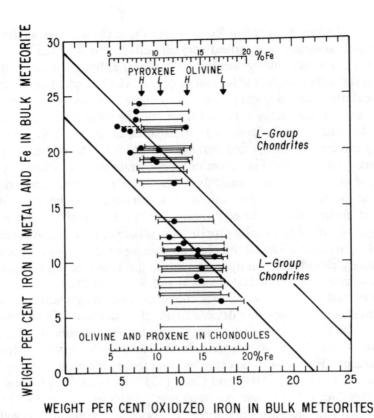

WEIGHT PER CENT OXIDIZED IRON IN BULK METEORITES

FIGURE 24 Urey-Craig diagram of metallic and sulfidic iron as a function of oxidized iron in the silicates. Each of the horizontal lines represents data from one meteorite; their end points denote values for the FeO content of olivine (right) and pyroxene (left) as determined by Keil and Fredriksson (1962) by an X-ray electron microprobe. These values are plotted on a 20% smaller scale for easier comparison. If Prior's Rule were valid, these points would lie on lines parallel to the diagonal lines. However, they lie almost vertically above each other. The solid circles denote the bulk FeO content, which does appear to follow Prior's Rule for each meteorite, but they do not agree with the mineral values and the fact that most of the silicate phase consists of the two minerals. (After Suess, 1964.) The reason for this discrepancy is not known (K. Keil, private communication).

contains somewhat more FeO than the pyroxene [metasilicate, $(MgO)_{1-x}(FeO)_x \cdot SiO_2$]. The individual chondrules in a meteorite contain these minerals in varying amounts; their melts, therefore, must have contained MgO and FeO in varying proportions. Yet, in the majority of ordinary chondrites, all the olivines have nearly the same FeO content and so have the pyroxenes (Keil and Fredriksson, 1964). To explain this apparent paradoxic, the assumption was made that the FeO originally present in the interstitial matter that had condensed in the solar nebula (at temperatures below the melting points of the silicates) was distributed into the magnesium minerals by diffusion during metamorphic processes. Chondrites with nearly constant MgO/(MgO + FeO) ratios were therefore, and still are, called "equilibrated chondrites". Chondrites with varying FeO content of their minerals are called "unequilibrated". This nomenclature may be convenient—it is now common usage—but it implies that the unequilibrated chondrites are the more primitive objects, that became "equilibrated" after they had formed. This is certainly not the case (Suess and Wänke, 1967).

It is not possible here to discuss the extensive literature that now exists on the subject, although a final explanation is extremely important for our ideas about the origin of the solar system. The explanation is very simple if we are willing to give up the contention that the chondrules formed in some way directly as condensation products of the primordial solar nebula: The chondrules may simply be considered to be molten grains of a preexisting igneous rock consisting of condensed solar matter, as suggested by Suess and Wänke (1967). In order to explain the differences in the composition of the different groups of ordinary chondrites, it might then be assumed that the three groups—the H, L, and LL groups—evolved from parent materials that had condensed at somewhat different oxygen fugacities. The parent body of the LL group appears to be the most oxidized, the L group somewhat less, and the H group the least, containing the most metal.

These differences in the degree of oxidation can be explained by the loss of hydrogen from the planetary nebula in such a way that, at the time of condensate formation in the area of the planetary disks, the loss was relatively low close to the Sun and greater farther from the Sun. Interestingly, the differences in the oxygen fugacities that must have prevailed during condensation are also reflected in the degree of mass-independent fractionation of the three oxygen isotopes (see Sect. 3.1 and Fig. 25).

The LL-group chondrites are frequently unequilibrated; that is, their chondrules show varying degrees of oxidation. They were for a long time

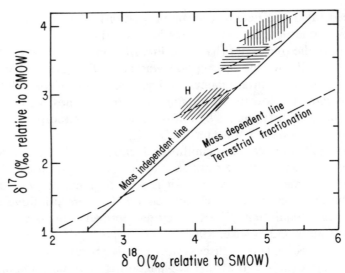

FIGURE 25 Isotopic composition of oxygen in ordinary chondrites (according to Clayton et al., 1976). The ratio $^{17}O/^{16}O$ is plotted against $^{18}O/^{16}O$ in per mill deviations from standard mean ocean water (SMOW). All terrestrial oxygen values fall along the broken line. Oxygen on the mass-independent solid line is obtained by laboratory experiments (see Fig. 19). The shaded areas indicate the ranges of oxygen values of bulk samples of H, L, and LL chondrites. Individual chondrules frequently give values that deviate considerably from these ranges (R. N. Clayton, personal communication).

erroneously considered to be the most primitive ordinary chondrites. Therefore, the unequilibrated chondrites have been more extensively investigated, and many features of the equilibrated chondrites are still poorly known. Thus, the iron content of the interstitial material compared with that of the chondrules has not been accurately analyzed. Further work on equilibrated chondrites would be highly desirable.

4.1.4 Enstatite Chondrites

The enstatite chondrites obtained their name from a mineral that consists of magnesium metasilicate, $MgSiO_3$. (It corresponds to the ferromagnesium silicate pyroxene, but without any iron.) They consist of material that was heated to high temperatures of some 1800°K or more at extremely low ambient oxygen fugacities. Whereas in carbonaceous chondrites practically all the iron is oxidized and present in oxides and silicates, the enstatite chondrites contain practically no oxidized iron; all the iron is present as metal. This is exactly what would be expected from substances that condensed directly from a hot gas of solar composition. It also corresponds to the relative amounts of magnesium and silicon in solar matter: $[Mg]/[Si] \simeq 1.07$.

At the gas pressure at which condensation is assumed to have taken place, that is, on the order of 10^{-3} atm, the oxygen fugacity is not low enough to account for the formation of nitrides, phosphides, sulfides, silicides, and so on, present in enstatite meteorites. It was therefore assumed that an equivalent amount of carbon, such as that present in carbonaceous chondrites, might have acted as the reducing medium that led to the formation of these unusual substances in enstatite chondrites. Indeed, some mineralogists have observed local indications of *in situ* reduction processes by elementary carbon.

Thermodynamic calculations by Herndon and Suess (1976) in which the possibility of pressures above 1 atm and corresponding condensation temperatures above 1800°K (i.e., above the triple point of iron), were considered showed that, together with refractory substances and liquid iron, all kinds of highly reduced unorthodox substances should form.

The first such mineral, discovered by British mineralogist Oldham (Story-Maskelyne, 1870), is CaS (calcium sulfide), called oldhamite. This substance reacts with water at room temperature and yields H_2S, with a smell characteristic of enstatite meteorites and some others. Another such unorthodox mineral is TiN, osbornite. For TiN formation according to the equation $TiN(s) + O_2(g) \rightleftharpoons TiO_2(s) + \frac{1}{2}N_2(g)$ in a gas of solar composition and 1 atm total pressure, Herndon gave a temperature of 1960°K at which TiN is stable. At 100 atm, TiN is stable above 1430°K. The formation of CaS is explained by the low volatility of this mineral as compared with CaO or SiO_2, which reacts with H_2 at high pressures to give SiO and finally SiH_4 in the gas phase. Because of these volatilities, the equilibrium $CaO + H_2S \rightleftharpoons CaS + H_2O$ will be moved to the right. The condensation of CaS should occur at a temperature above 1950°K and a total gas pressure above 10 atm. If the astronomers tell us that such high temperatures and pressures are inconceivable for the early planetary nebula, then we have to assume that the presence of carbon is responsible for the extremely reducing conditions. But can the reduction then have occurred in the gas phase, or after condensation, *in situ*? These questions are not easy to answer and require further study.

4.1.5 Achondrites, Stony Irons, and Iron Meteorites

Meteorites that have a chemical and mineralogical composition that does not reflect gas-phase reactions such as condensation or evaporation processes are frequently called "differentiated meteorites". As discussed in the previous sections, the chemical properties of the various types of chondrites indirectly or directly have to do with the condensation sequence of the chemical elements. In addition, the differentiated meteorites have evolved in a way similar to the terrestrial igneous rocks in magmatic processes, by the melting and cooling of magma.

In meteorites the most important of such differentiation processes is the separation of metallic nickel–iron from silicates and other slag-forming substances. This separation, however, may also have resulted from a difference in condensation temperatures, as, at sufficiently high temperatures, nickel–iron condenses before the silicates (see Sect. 3.2.1).

In some chondrites, as well as in stony irons, there are indications that liquid metal separated from silicates by settling out in a gravitational field. In some cases the gravitational field can be recognized by the way liquid troilite (FeS) separated from the metal. The troilite in irons is frequently found in pear-shaped inclusions that, while liquid, had moved "upwards", leaving a trail of solid impurities at the pointed end.

The stony irons are composed of a metal matrix in which ferromagnesium silicate crystals are embedded. The melting points of these crystals are well above that of the nickel–iron. Again an ambient gravitational field can be recognized, as the floating silicate crystals often tend to accumulate in a particular direction. What is especially important is the fact that the silicate usually contains oxidized iron; this means that these meteorites cannot have formed by direct condensation from a solar nebula, as in that case all the iron would be metallic. The metal in meteorites always contains nickel, in amounts that depend on the fraction of iron that has been oxidized.

The iron meteorites present interesting metallurgical problems. They are discussed in an extensive literature, and a three-volume work by Buchwald (1975) contains interesting details. Here, it must suffice to mention that nickel–iron occurs in two polymorphic forms: one body-centered cubic, called kamecite, or α-Ni-Fe; and a second face-centered cubic, called taenite, or β-Ni-Fe. These two crystalline forms give rise to the well-known "Widmanstätten figures" typical of meteoritic irons. These figures give information about cooling rates, and thus, on the size of the parent bodies from which the irons came.

Of particular interest are the achondrites, stony meteorites that do not contain any chondrules and no metal particles. They are in many ways similar in composition to the terrestrial basalts. It is now generally assumed that just as the material of the upper mantle of the Earth consists of whatever remained after subtraction of the siderophile elements from meteoritic matter, so the achondrites consist largely of what was left after the nickel–iron with the accompanying siderophile elements had accumulated in the center of some parent body.

Usually seven types of achondrites are listed. Some of these, however, are closely related and assumed to derive from one and the same parent body, presumably an asteroid, satellite, or planet. Closest to the composition of the basalts of the Earth and the Moon are the eucrites and the closely related diogenites, howardites, and aubrites. These achondrites have Rb/Sr ages of 4.5×10^9 years, the same as most meteorites. The second group consists of the shergottites and the chemically related nakhlites and chassignites, which exhibit much younger Rb/Sr ages of less than 3×10^9 years. These two groups of achondrites differ more than any type of chondrite from the primordial C-1 composition.

Achondrites are rare; they seem to come from the interior of large parent bodies, some of them possibly from the Moon or even the planets. The vast majority of stony meteorites, the chondrites, must come from the surfaces of relatively large objects.

4.2 ASTEROIDS AND COMETS

The distance from the orbit of Mars to the orbit of Jupiter is about twice that expected from Bode's law, an empirical expression describing the approximate radii of the orbits of planets as a function of the distance from the Sun. Instead of the planet that should exist according to Bode's law, a large number of small objects were found orbiting the Sun between Mars and Jupiter. In searching for the missing planet, the first object, Ceres, was discovered in 1801 by the Italian astronomer Piazzi. Its mass, however, was found to be less than 1/1000th that of the Earth. It has a diameter of only about 1000 km, and its density, about 5 g/cm^3, corresponds roughly to that of meteorites. The total mass of all the objects that populate the space between Mars and Jupiter is estimated to be less than 0.1% that of the Earth, or less than 10^{25} g. Many possible explanations have been advanced regarding the absence of a normal planet. Alfven and Arrhenius (1976) have invoked electromagnetic forces that prevented its formation. I prefer the idea that the absence of a planet has to do with the volatility of H_2O. As was repeatedly pointed out by Urey (1952), assuming the present solar luminosity, H_2O has an appreciable vapor pressure at distances closer to the Sun than the asteroidal belt, but has practically no vapor pressure, and can be considered rock-forming at the temperatures prevailing in the area of the orbit of Jupiter and beyond.

We have no accurate way to determine the luminosity of the Sun before it underwent its so-called T-Tauri state of deuterium burning, a state invoked by astrophysicists to explain many details in the formation of the solar system. In fact, much data, especially from the petrology of meteorites, indicate that some kind of a hot spell must have occurred early in the life of our solar system. It may be that relatively large amounts of ^{26}Al formed in nuclear reactions during the gravitational collapse and formation of the solar nebula and the Sun. ^{26}Al (half-life, 7×10^5 years) may have led, through its radioactive heating, to temperatures above the triple point and perhaps to complete evaporation of the condensed matter in large objects in the vicinity of the Sun (Wänke, 1981). Whipple (1976) and Cameron (1966) discussed the possibility of electric discharges, *viz.*, lightning, as the source of the melting. Certainly the work of Sonett (1979) should be mentioned here. Sonett investigated various possibilities of how the Sun, during a superluminous state, could have caused the melting of such small objects as the protochondrular grains.

In any case, it seems plausible to assume that conditions were right for the formation of objects consisting of rock-forming material (metals, silicates, etc.) in the inner part of the solar system and of objects containing ices (H_2O, NH_3) in its outer part (see Sect. 4.2). Then, it may well be that, at an intermediate distance from the Sun, conditions did not favor the preservation of either kind of object for longer periods. Instead of a single large planet, a large number of relatively small ice-

time when Jupiter's satellite system formed, because the densities of the Galilean satellites decrease markedly with increasing distance from the planet (Table XII), and hence the ratio of the ices to the rocks (class II to class I substances) increases with increasing distance from Jupiter. This agrees with observations of the surface of Io that indicate some kind of volcanic activity. The case of Saturn appears to be more complicated. The density of Saturn's Rea indicates that the satellite may consist of ices alone.

Saturn and Uranus possess ring systems in addition to their satellites. These rings are of great interest in connection with the mechanics of satellite formation. The rings consist of solid objects, of centimeter size in the case of the Saturn and considerably larger in the case of Uranus. The Saturnian rings show a spectrum of reflected sunlight similar to that of ice, but the presence of rocky substances can also be anticipated. The rings of Uranus have a conspicuously low albedo that indicates the presence of a black substance, presumably carbon. The presence of higher hydrocarbons can be anticipated also. Ionizing radiation should produce polymers and carbon black.

The Jovian planets have massive, deep, and turbulent atmospheres that contain H_2, He, NH_3, CH_4, and many other volatile substances. NH_3 and CH_4 lines are observed in the spectra of all the Jovian planets. On Jupiter and Saturn these gases are present in nearly solar proportions; this seems also to be the case for Uranus and Neptune. A serious difficulty in quantitative evaluation of the observations is the fact that the solar abundance of hydrogen, and thus also of helium, is less accurately known than the abundances of all the other pertinent elements. This is because the [H]:[Si] ratio is difficult to determine, the uncertainty being perhaps as much as a factor of two. Therefore, the important ratio of class I to class II elements is not known accurately, and the anticipated ratio of ices to rocks in Jovian satellites, as it should follow from the elemental abundance values, cannot be determined. Recent astronomical values for solar [H]:[Si] (Table V) indicate a ratio of ices to rocks of only slightly greater than one.

An interesting problem is presented by the colors of the clouds in the lower atmospheres of Jupiter and Saturn. They range from deep red to brown and yellow. Many substances have been suggested as being responsible for these colors. To chemists they are reminiscent of the polysulfides precipitated in aqueous solutions and also of the substances that form from SiO to give SiO_2 and elementary silicon. Both polysulfides and SiO should be present in large amounts at some depths in the atmosphere of Jupiter. Somehow, however, these substances supposedly do not fit into the model atmospheres considered. Photochemical reactions may have produced suitable substances.

The hydrides of nitrogen (NH_3), carbon (CH_4), phosphorus (PH_3), sulfur (H_2S), and other elements are present in the form of clouds and as gases in amounts corresponding to their vapor pressures at the respective temperatures. NH_3 clouds are observed on all the outer planets and on the giant Saturnian moon Titan (Wildt, 1934). The clouds give information on prevailing temperatures.

4.3.3 The Inner or Terrestrial Planets

Certainly more is known about our neighbors in the solar system—the planets Mars and Venus and the three moons, including our own—than about any other part of the solar system. Space probes have visited these planets, have landed, and have taken photographs and transmitted data on many kinds of measurements. Humans have been on our Moon and have returned samples of surface material. These endeavors have been disappointing for some, who had hoped to learn about extraterrestrial life or about some other planet that could be colonized. But the surface of Venus, our sister planet, is at a temperature of about 700°K, glowing dull red-hot, and is covered by an atmosphere of almost pure CO_2 at some 90 atm. The surface of Mars does not seem to be much more hospitable. Temperatures are an extremely cold 200°K, and only during Martian summer may they rise to the melting point of ice in subsolar regions. The atmosphere consists mainly of CO_2, and its surface pressure is about 6 mbar. This is slightly above the triple point pressure of water, so that liquid H_2O could exist under favorable conditions on the surface. The pressure, however, is considerably lower than the vapor pressure of human blood at normal human temperature, and without a spacesuit on Mars, human blood would boil immediately.

Why are the atmospheres of our neighboring planets so very different from that of our Earth, despite the fact that many other physical properties are quite similar? As we shall see, only slight differences in the geophysical parameters could have made our Earth equally uninhabitable. Even now, we have the ways and means to make life on Earth permanently impossible.

I cannot review here the whole field of geochemistry and its relationship to the chemistry of the other terrestrial planets. Only a few basic, and still in part unsolved, questions will be briefly discussed. The last few decades have brought about a firm conviction that we know the chemical composition of the matter from which the solar system and the Earth were formed. This is not a hypothesis, not a crude approximation. In the preceding discussion, an attempt has been made to demonstrate that our conclusions are firmly based on a number of different scientific fields, including nuclear physics, astronomy, geochemistry, and others.

The very large difference in the partial pressure of CO_2 on Mars, Venus, and Earth is relatively easy to explain: In thermodynamic equilibrium, this partial pressure results from the reactions of CO_2 with silicates, for example,

$$CO_2 + CaSiO_3 \rightleftharpoons CaCO_3 + SiO_2.$$

On the Earth, this and similar reactions are generally termed rock-weathering reactions, which proceed in aqueous solutions or on wet surfaces. Our atmosphere contains about 300 ppm of CO_2, corresponding roughly to an average of the silicate–carbonate equilibrium pressure at the average temperature on Earth. This was first pointed out by Urey (1952) and is called the "Urey equilibrium pressure". It is not surprising that this pressure is quite different on Venus. There, with no H_2O on its surface, it is not established at normal temperatures within a reasonable length of time.

It is interesting to note that the amount of CO_2 in the carbonates on Earth is on the same order of magnitude as the amount of CO_2 in the atmosphere of Venus. As early as 1940, Wildt had predicted a high temperature on the surface of Venus resulting from a greenhouse effect caused by a high partial pressure of CO_2 in the atmosphere (Wildt, 1940). Wildt's predictions have now been fully confirmed.

Compared to conditions on Venus the CO_2 pressure on Mars is low, only approximately 20 times that in the atmosphere of the Earth. On Mars, regional topographic indications of periods with liquid water H_2O can be recognized. The existence of carbonates can therefore be expected. The excess of CO_2 relative to the Urey equilibrium pressure is presumably due to volcanic exhalations.

In the very early stages of formation, Venus, Earth, and Mars must have had enough surface H_2O so that photolysis and subsequent hydrogen loss supplied sufficient oxygen for the oxidation of carbon and hydrocarbons. The very low neon content (Sect. 3.1.1.4) indicates that this H_2O was secondary—derived from exhalations after the primordial atmospheres were lost.

Space probe pictures show in an impressive way that all objects in our solar system with surfaces not protected by gases or liquids bear the effects of bombardments by large bodies in the form of impact craters. Pictures of our Moon taken nearly a century ago and modern pictures of Mercury look almost indistinguishable to the nonexpert.

Obviously, there is every reason to suspect that the average composition of the inner planets corresponds to the average composition of the stony meteorites, even though surface rocks of Earth and Moon show considerable deviations in the content of many elements. By analogy with meteorites, the existence of an iron core of the Earth had been suspected in the late 19th century. Then, analyses of earthquake wave propagation showed that a discontinuity existed (Wiechert, 1897) some 1500 km below the planet's surface. This discontinuity was assumed to be the boundary between a stony mantel and a nickel–iron core. The modern view of the internal structure of the Earth is depicted in Figure 26.

There is reason to believe that the structures of the other terrestrial planets are basically similar, except in one regard: Differences in the ratios of silicates to metal should account for the differences in densities (Table XI). Most probably, the planet Mercury contains a larger proportion of metallic iron and the planet Mars a somewhat smaller one

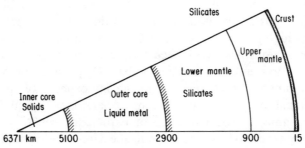

FIGURE 26 Currently accepted cross section of the interior of the Earth (see, for example, Ringwood, 1979).

than the Earth. The Moon may have a small iron core or none at all. In any case, some process that separated the metal from the silicates must have occurred when these objects formed.

The nature of this process is still an object of considerable controversy. The most obvious and simplest assumption is that the metal settled down through the much lighter silicates in the gravitational field of a large body, in a way similar to magmatic separation on the surface of the Earth. However, an argument by Kuhn and Rittman (1941), now considered erroneous, maintained that the viscosity of molten silicates and molten iron at the pressures prevailing in the Earth's interior is so large that a settling down even over geological periods of time would have been impossible. Kuhn and Rittman concluded that an iron core could not have formed and, hence, that the Earth's interior should consist mainly of highly compressed, dense hydrogen. Several ideas to overcome this apparent dilemma (see, for example, Elsasser, 1963) have been suggested; the most straightforward is the one proposed by Eucken (1944), who pointed out that successive condensation, not magmatic separation, may have led to an iron core. We still do not know with certainty which of these processes, magmatic separation or fractionated condensation, predominated.

The results of a very large number of chemical analyses of surface rocks from the Earth and Moon, as well as of meteorites, can now be compared with the solar abundances of the elements and the C-1 data. If an element is deficient in the silicate phase, that is, if its content is lower than expected from its solar abundance value, then the missing fraction of this element must be present elsewhere. Either this element because of its volatility, did not fully condense with the bulk of the material or, because of its chalcophile or siderophile tendencies was taken up and incorporated in another phase, primarily in the metal. For a chemist it is easy to see that the lower the oxygen fugacity during condensation, namely, the more reducing the conditions, the more siderophile the elements will behave. This was shown by Wänke and his coworkers (Wänke et al., 1981) by determining the distribution coefficients of several metals (such as gallium, cobalt, nickel, tungsten) between silicates and nickel–iron as a function of oxygen fugacity. Wänke and Dreibus (1979) argued that the deficiency by more than a factor of two of vanadium, chromium, and manganese on the Earth and Moon and in the eucrite parent body can hardly have been caused by volatility. More likely, these elements went into the nickel–iron core as sulfides or metal

(Dreibus and Wänke, 1984). Similar criteria can be applied to other deficient elements. The important element silicon is slightly deficient (by about 6%) on Earth and Moon and in some achondrites. This was explained by the high silicon content of the Earth's core (Ringwood, 1978) and, in particular, by the solid inner core that, according to Herndon (1979), consists essentially of nickel silicide (Ni_2Si).

These and similar considerations have now led to the following picture of the formation of the Earth and the other inner planets. The analytical data obtained at Mainz by Wänke and his coworkers agree with the assumption of Ringwood (1979) that accretion of the inner planets started with highly reduced material. Wänke considered two components: Component A is free of volatile and moderately volatile elements. Chromium, mangnanese, vanadium, and the siderophile elements are all present as metals or sulfides; even a portion of the silicon is in a metallic state. Because of the high temperatures reached during the accretion (Safranov, 1978; Kaula, 1979), segregation of metal, that is, core formation, occurred at the same time as accretion after the Earth had reached about 10% of its mass. When the Earth had reached about two-thirds of its mass, more and more of component B, which was fully oxidized and contained all the condensable elements in C-1 abundances, began to be added. Wänke estimated the mixing ratio of components A and B for the Earth to be 85 : 15.

Wänke's inhomogeneous accretion model accounts in a satisfactory way for the elemental composition of the Earth and in particular for the relatively high concentrations of nickel and cobalt in the Earth's mantle. Its main point is that the assumption of a highly reduced state at the beginning of accretion is unavoidable and that the metal of the Earth's core was never in chemical equilibrium with the Earth's mantle in its present form (Wänke, 1981). Remarkably, the abundances of nickel, cobalt, chromium, manganese, and vanadium are the same in the Earth's mantle as in the Moon. This observation provides a strong argument in favor of a common origin of the two bodies (Wänke and Dreibus, 1986).

FIVE

Concluding Remarks

I have organized the subject matter in this book differently than is commonly done, discussing first nuclei, then atoms, then chemical compounds, condensates, and finally the planets. Instead of describing the solar system in order of size, it could be described more conventionally, in a chronological order of evolution, perhaps as follows.

5.1 A POSSIBLE CHRONOLOGICAL ORDER OF COSMOCHEMICAL EVENTS

Element synthesis (Sect. 2.2) more than 7×10^9 years ago.

Mixing of the products, the R-, S-, and possibly other genetic components as very hot gases.

Cooling of the perfectly mixed gases and condensation of refractory dust. Forming of icy agglomerates such as are present today in the Oort cloud (Sect. 4.1.2).

Accretion into protoplanets, perhaps with cores consisting of nickel–iron.

Formation of objects such as are now in the *outer solar system*, that is, Jovian planets with rings and satellites (Sect. 4.3.2).

Hot spell or series of hot spells, T-Tauri state(s) of deuterium burning (Cameron, 1963), affecting mainly the *inner solar system*.

Disintegration of icy agglomerates, including those in the asteroid belt.

Partial re-evaporation of volatiles and perhaps moderately volatiles at temperatures exceeding 2000°K.

Accretion of planetary parent bodies about 4.6×10^9 years ago.

Component A: Metallic nickel–iron, highly refractory materials, then magnesium silicates (Sect. 4.3.3).

Beginning of hydrogen loss (Sect. 3.2.2); formation of ferromagnesium silicates, terrestrial mantle material.

Partial refragmentation of meteorite parent bodies in the asteroidal belt, reheating of debris, chondrule formation.

Component B: Fully oxidized material.

In the inner part of the solar system, loss of all gases and almost all H_2O.

Formation of the inner (terrestrial) planets, the Moon, and the two satellites of Mars.

This sequence of events suggested here is by no means final, although something like it must have occurred in about that order. We are well informed about some of the processes, but know little or nothing about others. For example, we have quantiative data on the physical parameters of the S-process in nuclear synthesis, but we know little about the R-process, except that it is possible to establish conjectures that may or may not be correct. We know that during the history of the matter that surrounds us the temperature must have varied greatly between the present temperature and the melting points of the silicates, around 1800°K. However, we know little about time and amplitude of these temperature variations. We do know that almost all the gases originally associated with our Earth and the other inner planets have left the inner part of the solar system, but we do not know at what stage and by what mechanism.

Not included here are the important models of the origin of the solar system that are based primarily on mechanical features, such as gravitational collapse of a rotating gas mass, or on the more recently investigated phenomena of magneto hydrodynamics, particle radiation, and plasma effects. This text has been limited to a discussion of the chemical behavior of matter. Chemical questions were largely neglected until Urey

(1952) emphasized chemical arguments in connection with astronomical problems.

A large part, perhaps too large a part, of this discussion has described elementary facts of nuclear physics. But how can a student understand the chemical composition of our cosmic environment without some understanding of ''magic numbers'' in nuclear structure. How can a student of cosmochemistry understand and interpret the chemical composition of the members of the solar system without knowing something about element synthesis and the origin of the elements?

Although many questions remain unanswered, a path to their solution can be discerned. The solutions to other questions, it can be safely said, are now firm scientific knowledge. This knowledge clearly shows how very unique our planet Earth is, with free oxygen in its atmosphere and liquid H_2O on its surface, features that, from a scientific viewpoint, must be quite rare in our universe.

References

Alfven, H. and Arrhenius, G., *Evolution of the Solar System*, NASA, Washington, D.C. (1976).

Aller, L. M., *The Abundance of the Elements*, Interscience, New York (1961).

Amiet, J. P., and Zeh, H. D., Z. Phys. **217,** 485 (1968).

Anders, E., and Ebihara, M., Geochim. Cosmochim. Acta **46,** 2363 (1982).

Anderson, D. L., Moon **8,** 33 (1973).

Arnold, J. R., and Suess, H. E., Annu. Rev. Phys. Chem. **20,** 293 (1969).

Aston, F. W., *Mass. Spectra and Isotopes*, London (1933).

Bainbridge, K. T., Phys. Rev. **37,** 1717 (1931).

Barth, C. A., and Suess, H. E., Z. Phys. **158,** 85 (1960).

Begemann F., in *Earth Science and Meteorites,* J. Geiss and E. Goldberg Eds., North-Holland Publ. Co., Amsterdam (1963)

Bigeleisen, J., and Mayer, M. G., J. Chem. Phys. **15,** 261 (1947).

Boato, C., Geochim. Cosmochim. Acta **6,** 209 (1954).

Boudouard O., Ann. Chim. Phys. (Pan's) Ser. 7. **24,** 5/85 (1901).

Brown. H., in *The Atmospheres of the Earth and Planets*, G. P. Kuiper, Ed., Univ. of Chicago Press, (1948), p. 258.

Brown, H., Revs. Modern Phys. **21,** 625 (1949).

Brown, H., J. Astrophys. 641 (1950).

Buchwald, V. F., *Handbook of Iron Meteorites*, Univ. of California Press, Berkeley (1975).

Burbidge, E. M., Burbidge, G. R., Fowler, W. A., and Hoyle, F., Rev. Mod. Phys. **29,** 547 (1957).

Cameron, A. G. W., Icarus **1,** 339 (1963).

Cameron, A. G. W., in *Origin and Distribution of the Elements*, L. H. Ahrens, Ed., Pergamon, New York (1968), p. 125.

Chadwick, J., Proc. Soc. London, Ser. A **136,** 692 (1932).

Clarke, F. W., Bull. Phil. Soc. Washington **11,** 131 (1889).

Clayton, R. N., and Mayeda, T. K., Geophys. Res. Lett. **4,** 295 (1977).

Clayton, R. N., Onuma, N., and Mayeda, T. K., Earth Planet. Sci. Lett. **30,** 10 (1976).

129

Dodd, R. T., *Meteorites*, Cambridge Univ. Press, Cambridge, U.K. (1981).

Dreibus, G., and Wänke, H., in Proceedings of the 27th International Geological Congress, Vol. 11 VNUC Science Press, Moskow (1984).

Eberhardt, P., Earth Planet. Sci. Lett. **24**, 182 (1974).

Elsasser, W., J. Phys. Rad. **4**, 549 (1933); **5**, 389, 635 (1934).

Elsasser, W., in *Earth Science and Meteoritics*, J. Geiss and E. Goldberg, Eds., North Holland Publ. Co. Amsterdam (1963).

Esat, T. M., Spear, K. Y., and Taylor, S. R., Lunar Planet. Sci., **16**, 210, (1985).

Eucken, A., Nachr. Acad. Wiss. Göttingen (1944).

Fahey, A. J., Goswami, J. N., McKeegan, K. D., Zinner, E., Lunar Planet. Sci. **16**, 299 (1985).

Farkas, A., and Farkas, L., Proc. Soc. London **115**, 373 (1934).

Faltings, V., and Harteck, P., Z. Naturforsch **5a**, 438 (1950).

Fersman, A. E., *Geochemice*, Acad. Nauk USSR, Leningrad (1934).

Flerov, G. N., and Petrzhak, K. A., Phys. Rev. **58**, 89 (1940).

Fowler, R. H., and Guggenheim, E. A., *Statistical Thermodynamics*, Cambridge University Press, Cambridge, U.K. (1939).

Friedlander, G., Kennedy, J. M., and Miller, J. M., *Nuclear and Radiochemistry*, 2nd ed., Wiley, New York (1966).

Gamow, G. Z., Physik **51**, 204 (1928).

Geiger, H., and Nuttall, J. M., Phil. Mag. **22**, 613 (1911); **23**, 439 (1912).

Geiss, J., and Reeves, H., Astronomy Astrophys. **93**, 189 (1981).

Gerling, E., and Lefskii, L. K., Dok 1. Acad. Nauk USSR **110**, 750 (1956).

Göbel, R., Ott, V., and Begemann, F., J. Geophys. Res. **83**, 855 (1978).

Goldschmidt, V. M., Nor. Vidensk. Akad. Oslo, Mat. Nat. Kl. **1937**, 4. Goldschmidt, V. M., *Geochemistry*, Oxford Press, London (1954).

Grossman, L., and Larimer, J. W., Rev. Geophys. Space Phys. **12**, 71 (1974).

Groth W., and Suess, H. E., Naturwissenschaften **26**, 77 (1938).

Hahn, O. and Strassmann, F., Naturwissenschaften **27**, 11 (1939).

Harkins, W. D., J. Am. Chem. Soc. **39**, 856 (1917).

Harkins, W. D., Chem. Rev. **5**, 371 (1928).

Harteck, P., and Jensen, H. J. D., Z. Naturforsch. **3a**, 591 (1948).

Harteck, P., and Suess, H. E., Naturwissenschaften **36**, 218 (1949).

Haxel, O., Jensen, H. J. D., and Suess, H. E., Naturwissenschaften **36**, 376 (1948).

Haxel, O., Jensen, H. J. D., and Suess, H. E. Z. Phys. **128**, 295 (1950).

Herndon, J. M., Naturwissenschaften **69**, 34 (1982).

Herndon, J. M., Proc. Roy. Soc. London, Ser. A. **363**, 283 (1978).

Herndon, J. M., Proc. Roy Soc. London, Ser. A. **372,** 149 (1980).

Herndon, J. M., Proc. Roy. Soc. London, Ser. A. **368,** 495 (1979).

Herndon, J. M., and Suess, H. E., Geochim. Cosmochim. Acta **40,** 395 (1976).

Hey, M. H., *Catalogue of Meteorites*, British Museum, London (1966).

Hughes, D. J., and Sherman, D., Phys. Rev. **78,** 632 (1950).

Hunten, D. M., J. Atmos. Sci. **30,** 1481 (1973).

Jeans, J., *Astronomy and Cosmogony*, Cambridge Univ. Press, London (1928).

Jensen, J. H. D., and Suess, H. E., Naturwissenchaften **32,** 374 (1944).

Käppeler, F., Beer, H., Wisshak, K., Clayton, D. D., Macklin, R. L., and Ward, R. A., Astrophys. J. **257,** 821 (1982).

Kaula, W. M., J. Geophys. Res. **84,** 999 (1979).

Keil K. and Fredriksson K., J. Geophys. Res. (1964)

Kuhn, W., and Rittman, A., Geol. Rundsch. **32,** 215 (1941).

Kuroda, P., *The Origin of the Chemical Elements and the Oklo Phenomenon*, Springer-Verlag, New York (1982).

Lewis, J. S., and Prinn, R. G., *Planets and Their Atmospheres*, Academic, New York (1984).

Lee, T., Papanastassiou, D. A., and Wasserburg, G. J., J. Geophys. Res. **3,** 41 (1976).

Libby, W. F., Phys. Rev. **55,** 1269 (1939).

Martell, E. A., and Libby, W. F., Phys. Rev. **80,** 977 (1950).

Marti, K., Science **166,** 1263 (1969).

Marti, K., Wilkening, L. L., and Suess, H. E., Astrophys. J. **173,** 445 (1972).

Mayer, M. G., Phys. Rev. **74,** 235 (1948).

Mayer, M. G., Phys. Rev. **78,** 16 (1950).

Mayer, M. G., and Jensen, H. J. D., *Elementary Theory of Nuclear Shell Structure*, Wiley, New York (1955).

Mayer, M. G., and Teller, E., Phys. Rev. **76,** 1226 (1949).

Mattauch, J., Z. Phys. **91,** 361 (1934).

Mattauch, J., and Herzog, R., Z. Phys. **89,** 786 (1934).

McCord, T. B., and Gaffey, M. J., Science **186,** 352 (1974).

Menzel, D. H., Pub. Lick Observ. **17,** (1931).

Merryhue, C. M., Pepin, R. O., and Reynolds, J. H., J. Geophys. Res. **67,** 2017 (1962).

Miller, S. L., and Urey, H. C., Science **130,** 245 (1959).

Noddack, I., Angew. Chem. **47,** 835 (1936).

Noddack, I., and W., Naturwissenschaften **35,** 59 (1930).

Oort, J. H., Bull. Astron. Inst. Netherlands **11,** 408 (1950).

Palme, H., Suess, H. E., and Zeh, H. D., *Landoldt-Börnstein* VI-2A, 257 (1981).

Partsch P., *Die Meteorite oder vom Himmel gefallene Steine und Eisenmassen*. Wien, Verl. Kaulfuss Wittwe (1843).

Payn C. H., in *Stellar Atmospheres*, Harvard Observatory, Monograph No. 1, Cambridge, Mass. (1925).

Perlman, I., Ghiorso, A., and Seaborg, G. T., Phys. Rev. **77**, 26 (1950).

Pillinger, C. T., Geochim. Cosmochim. Acta **48**, 2739 (1984).

Poole, J. H. J., Sci. Proc. Dublin Soc. **22**, 345 (1941).

Prior, G. T. and Hey, M. H., *Catalogue of Meteorites*, Transfers of the British Museum, London (1953).

Reynolds, J., J. Geophys. Res. **68**, 2939 (1963).

Ringwood, A. E., Geochim. Cosmochim. Acta **15**, 195 (1979).

Ringwood, A. E., *Origin of the Earth and the Moon*, Springer-Verlag, New York (1979).

Russel, H. N., Astrophys. J. **70**, 11 (1929).

Safronov, V. S., Icarus **33**, 3 (1978).

Schmidt, T., Z. Phys. **106**, 358 (1937).

Sears, D. W., *The Nature and Origin of Meteorites*, Oxford Univ. Press, New York (1978).

Senftle, F. E., Stieff, L., Cuttitta, F., and Kuroda, P. K., Geochim. Cosmochim. Acta **11**, 189 (1957).

Shima, M., Geochim. Cosmochim. Acta **50**, 577 (1985).

Signer, P., and Suess, H. E., in *Earth Science and Meteoritics*, J. Geiss and E. Goldberg, Eds., North-Holland Publ. Co., Amsterdam (1963), p. 241.

Sonett, C. P., Geophys. Res. Lett. **6**, 677 (1979).

Spitzer, L., Jr., in *The Atmospheres of the Earth and Planets*, C. P. Kuiper, Ed., Univ. of Chicago Press, Chicago, (1952) p. 211.

Story-Maskelyne, N., Phil. Trans. Soc., London **160**, 195 (1870).

Ströomgren, B., Festschr. E. Strömgren (1940).

Suess, Eduard, in *Das Antlitz der Erde*, Vol. 3, Pt. 2, p. 626, F. Tempsky and E. Freitag, Wien and Leipzig (1909).

Suess, H. E., Z. Naturforsch. Teil A **2a**, 311, 604 (1947).

Suess, H. E., Experientia **5**, 266 (1949a).

Suess, H. E., Z. Naturforsch. Teil A **4a**, 328 (1949b).

Suess, H. E., J. Geol. **57**, 237 (1949c).

Suess, H. E., Trans. Am. Geophys. Union **34**, 343 (1953).

Suess, H. E., J. Geophys. Research **67**, 2029–2034 (1962).

Suess, H. E., in *Isotopic and Cosmic Chemistry*, p. 385, H. Craig, S. Miller, and G. Wasserburg, Eds., North-Holland Publ. Co., Amsterdam (1964).

Suess, H. E., in *Origin of the Solar System*, p. 143, R. Jastrow and A. G. W. Cameron, Eds., Academic Press, New York, (1968).

Suess, H. E., and Jensen, H. J. D., Arkiv Fysik 3, 577 (1951).

Suess, H. E., and Thompson, W. B., in *Chondrules and Their Origin*, E. A. King, Ed., Lunar and Planet. Institute, Houston (1983).

Suess, H. E., and Urey, H. C., Rev. Mod. Physics, **28**, 53 (1956).

Suess, H. E., and Wänke, H., Prog. Oceanog. 3, 347 (1965).

Suess, H. E., and Wänke, H., J. Geophys. Res. **72**, 3609 (1967).

Suess, H. E., Haxel, O., and Jensen, H. J. D., Naturwissenschaften 36, 153 (1949).

Tammann, G., Z. anorg. allgem. Chem., **131**, 96 (1923).

Thiemens, M. H., and Heidenreich III, J. E., Science **219**, 1073 (1983).

Thomson, J. J., Phil. Mag. **11**, 769 (1906).

Ubbelohde, A. R., Proc. Phys. Soc. **59**, 12; **61**, 96 (1948).

Unsöld, A., Z. Astrophys. **8**, 225 (1934); **23**, 75, 100 (1944).

Unsöld, A., Trans. Int. Astron. Union 7, 463 (1948).

Unsöld, A., Science **163**, 1015 (1969).

Unsöld, A., *The New Cosmos*, Springer-Verlag, New York (1977).

Urey, H. C., *The Planets, Their Origin and Development*, Yale Univ. Press, New Haven, CT (1952).

Urey, H. C., and Bradley, C. A., Jr., Phys. Rev. **37**, 843 (1931).

Urey, H. C. and Craig, H., Geochim. Cosmochim. Acta. **4**, 36–82 (1953).

Wänke, H., Phil. Trans. R. Soc. London, Ser. A **303**, 287 (1981).

Wänke, H., and Dreibus, G., in *Origin of the Moon*, Hartmann, Phillips, and Raylor, Ed., Lunar and Planetary Institute, Houston (1986).

Wänke, H., and Dreibus, G. in *Origin and Distribution of the Elements*, L. H. Ahrens., Ed., pp 99–108, Pergamon Press, Oxford 1979.

Wänke, H., Dreibus, G., Jagoutz, E., Palme, H., and Rammensee, W., Lunar Planet. Sci. **12**, 1139 (1981).

Wasserburg, G. T., Lee, T., and Papanastassiou, D. A., Geophys. Res. Lett. **4**, 299 (1977).

Wasson, J. T., *Meteorites, Their Record of Early Solar System History*, W. F. Freeman and Co., New York (1985).

Whipple, F. L., Astrophys. J. **111**, 375 (1950).

Whipple, F. L., Mem. Soc. Sci. Liege Collect. **9**, 101 (1976).

Wiechert, E., Nachr. Ges. Wiss., Göttingen, p. 221 (1897).

Wildt, R., Nature (London) **134**, 418 (1934).

Wildt, R., Astrophys. J. **91**, 266 (1940).

Author Index

Geiss, J., 85, *130*
Gerling, E., 87, *130*
Ghiorso, A., *132*
Göbel, R., 90, *130*
Goldschmidt, V. M., 5, 22, 49, 50, 54, 105, *130*
Goswami, J. N., 84, *130*
Grossman, L., 91, *130*
Groth, W., 97, *130*
Guggenheim, E. A., 77, *130*

Hahn, O., *130*
Harkins, W. D., 15, 43, *130*
Harteck, P., 78, 94, 95, *130*
Haxel, O., 22, 31, 32, *133*
Heidenreich, J. E., 82, 83, *133*
Herndon, J. M., 107, 111, 124, *130*, *131*
Herzog, R., 13, *131*
Hey, M. H., 103, *131*, *132*
Hoyle, F., 65, *129*
Hughes, D. J., 21, *131*
Hunten, D. M., 94, *131*

Jagoutz, E., 123, 124, *133*
Jeans, J., 94, *131*
Jensen, J. H. D., 22, 29, 31, 38, 56, 94, *130*, *131*, *133*

Käppeler, F., 68, *131*
Kaula, W. M., 124, *131*
Keil, K., 107, 108, 109, *131*
Kennedy, J. M., 14, 18, *130*
Kuhn, W., 123, *131*
Kuroda, P., 42, *131*, *132*

Larimer, J. W., 91, *130*
Lee, T., 82, *131*, *133*
Lefskii, L. K., 87, *130*
Lewis, J. S., 94, *131*
Libby, W. F., 44, *131*
Lugmair, 84

McCord, T. B., 115, *131*
McKeegan, K. D., 84, *130*
Macklin, R. L., 68, *131*
Martell, E. A., 43, *131*

Marti, K., 88, *131*
Mattauch, J., 15, *131*
Mayeda, T. K., 82, 110, *129*
Mayer, L., 11
Mayer, M., 30
Mayer, M. G., 22, 29, 65, 79, *129*, *131*
Mendeleev, D. I., 11
Menzel, D. H., 48, *131*
Merryhue, C. M., 87, *131*
Miller, J. M., 14, 18, *130*, *131*
Miller, S. L., 97, 101, *131*

Noddack, I., 40, 105, 106, *131*
Nuttall, J. M., 39, *130*

Onuma, N., 82, 110, *129*
Oort, J. H., 116, *131*
Ott, V., 90, *130*

Palme, H., 50, 100, 123, 124, *132*, *133*
Papanastassiou, D. A., 82, *131*, *133*
Partsch, P., 103, *132*
Payn, C. H., 48, *132*
Pepin, R. O., 87, *131*
Perlman, I., 40, *132*
Petrzak, K. A., 42
Pillinger, C. T., 82, *132*
Poole, J. H. J., 94, *132*
Prinn, R. G., 94
Prinz, R. G., *131*
Prior, G. T., 103, 107, *132*

Rammensee, W., 123, 124, *133*
Reeves, H., 85, *130*
Reynolds, J., 89, *132*
Reynolds, J. H., 87, *131*
Ringwood, A. E., 124, *132*
Rittman, A., 123, *131*
Russell, H. N., 48, *132*
Rutherford, 13, 15

Safranov, V. S., 124, *132*
Schmidt, T., 28, *132*
Seaborg, G. T., *132*
Sears, D. W., *132*
Senftle, F. E., 42, *132*
Sherman, D., 21, 131
Shima, M., 84, *132*
Signer, P., 87, 89, *132*

Subject Index

DATE DUE

MAR 2 1999	

GAYLORD   PRINTED IN U.S.A.